Routledge Revivals

State Mineral Enterprises

State ownership in mineral industries has increased massively from the 1950s affecting the world mineral sector greatly. Originally published in 1985, this study analyses the effects this had on the international market covering topics such as state takeovers of mineral firms, price stabilisation methods, state-owned enterprises in developing countries and whether state ownership will negatively impact private multinational companies. This title will be of interest to students of environmental studies.

State Mineral Enterprises

An Investigation into their Impact on International Mineral Markets

Marian Radetzki

First published in 1985
by Resources for the Future, Inc.

This edition first published in 2016 by Routledge
2 Park Square, Milton Park, Abingdon, Oxon, OX14 4RN
and by Routledge
711 Third Avenue, New York, NY 10017

Routledge is an imprint of the Taylor & Francis Group, an informa business

Publisher's Note
The publisher has gone to great lengths to ensure the quality of this reprint but points out that some imperfections in the original copies may be apparent.

Disclaimer
The publisher has made every effort to trace copyright holders and welcomes correspondence from those they have been unable to contact.

A Library of Congress record exists under LC control number: 85002346

ISBN 13: 978-1-138-12000-6 (hbk)
ISBN 13: 978-1-315-65187-3 (ebk)

State Mineral Enterprises

State Mineral Enterprises

An Investigation into Their Impact on International Mineral Markets

MARIAN RADETZKI

Resources for the Future
and
The Pennsylvania State University
in cooperation with
The International Institute for Applied Systems Analysis

Resources for the Future/Washington, D.C.

Manufactured in the United States of America

Published by Resources for the Future, Inc.
1616 P Street, N.W., Washington, D.C. 20036
Resources for the Future books are distributed worldwide by
The Johns Hopkins University Press

Library of Congress Cataloging in Publication Data

Radetzki, Marian.
State mineral enterprises.

Bibliography:.p.
Includes index.
1. Mineral industries—Government ownership—Developing countries—Case studies.
2. Mineral industries. 3. Competition, International. I. Title.
HD9506.D452R34 1985 333.8'5 85-2346
ISBN 0-915707-16-0

Resources for the Future is a nonprofit organization for research and education in the development, conservation, and use of natural resources, including the quality of the environment. It was established in 1952 with the cooperation of the Ford Foundation. Grants for research are accepted from government and private sources only on the condition that RFF shall be solely responsible for the conduct of the research and free to make its results available to the public. Most of the work of Resources for the Future is carried out by its resident staff; part is supported by grants to universities and other nonprofit organizations. Unless otherwise stated, interpretations and conclusions in RFF publications are those of the authors; the organization takes responsibility for the selection of significant subjects for study, the competence of the researchers, and their freedom of inquiry.

This book is a product of the Energy and Materials Division, directed by Joel Darmstadter. It forms a part of the joint RFF-Pennsylvania State University Mineral Economics and Policy Program. It was designed by Elsa B. Williams and edited by Charlene Semer. The index was prepared by Lorraine and Mark Anderson.

Contents

PART II. THE CASE STUDIES

PART III. CONCLUSIONS

TABLES

FIGURES

Preface

During the course of this study, I have benefited greatly from discussions with many professional colleagues who read and commented upon successive versions of the manuscript. I want to mention especially Phillip Crowson, Charles Kindleberger, Walter Labys, Hans Landsberg, Raymond Mikesell, Erik Moberg, Peter Svedberg, John Tilton, Raymond Vernon, two anonymous referees attached to the RFF Publications Committee, and the participants in a workshop convened by the International Institute for Applied Systems Analysis in Austria on October 17–19, 1983, to discuss a draft version of the present work. The many comments and criticisms I received have considerably improved the end product.

Special thanks are also owed to the managements of PT Timah in Indonesia, Ferrominera in Venezuela, and ZCCM in Zambia for their cooperation and hospitality during my work on the case studies of these three corporations.

The Institute for International Economic Studies at the University of Stockholm, my professional home for more than a decade, provided excellent physical and intellectual facilities for my work. The Swedish Agency for Research Cooperation with Developing Countries supplied a generous grant in support of this research project.

Marian Radetzki

Stockholm
1984

I

THE ISSUES

1

Introduction

During the past few decades, state ownership of developing countries' metallic mineral industries has undergone sizable expansion. In 1950, government equity holdings in these industries were insignificant; by 1980, such holdings constituted about half the metallic mineral production capacity in the Third World. Most of the output of these industries is exported. The purpose of the present book is to explore how the new owner group influences and changes the international metallic mineral markets.

This book is about the public ownership of metallic minerals; mineral fuels and nonmetallics, such as petroleum, coal, asbestos, and fluorspar, are not considered. In what follows, the word "minerals" refers to metallic minerals only.

Although public ownership in the mineral industries is in no way limited to the Third World, the study places heavy emphasis on state enterprises in developing countries (LDCs). This emphasis is partly justified by the faster growth and higher proportion of state ownership in the Third World compared to developed countries. Another justification is the special concern of the international mining industry about the repercussions likely to arise from LDC nationalizations. These actions have often been taken by economically inexperienced governments, in ventures that have been very large in relation to the national economy.

The central theme of the book is how the emerging state involvement in the mineral industries affects the functioning of international mineral markets. The goals and behavioral characteristics of state-owned mineral firms are studied with a view to clarifying how these firms might affect such market variables as costs and prices, competitive conditions, vertical integration, or the overall volume and geographical allocation of investments in new production capacity.

For some time, the rapid growth of state ownership of developing countries' mineral production by nationalizing existing capacity and participating in new ventures has been apparent. Actual information on the total extent of state involvement has at best been fragmentary, however. This book is a response to the need for an overall picture of the state enterprise phenomenon. It contains a detailed discussion of the various measures that could be used to quantify state ownership and includes data sets for the major mineral industries, to which the measures are then applied. The book also describes how current governmental ownership positions evolved in the past and conjectures about their likely significance in the future.

A more important reason for the book, however, is to address the increasing concerns about the implications of growing state ownership in developing countries' mineral industries. These concerns have seldom been thought through analytically, and they point in opposite directions.

One important worry has been that nationalization of foreign-owned ventures in developing countries on a large scale will result in inadequate mineral supply and rising mineral price levels, with harmful consequences for user industries in importing countries.[1] The underlying argument is that state-owned enterprises are so inefficient and so heavily taxed by their owner governments that their profits are not sufficient to finance exploration, capacity expansion, or even maintenance of existing capacity.[2] Under these conditions, supply tends to contract, pushing prices up, at least temporarily.[3]

An opposite concern has been that the actions of state mineral enterprises in developing countries will result in excess supply and lower average price levels, with severely detrimental consequences to privately owned mining industries. The proponents of this view point to either or both of two factors. First, when governments assume control, the national ambition to speed up economic progress takes myopic precedence over careful analyses of market prospects, and mineral deposits are developed at a faster pace than they would have been under a private multinational regime. The result is excess capacity

1. R. Mikesell, *New Patterns of World Mineral Development* (London, British North America Committee, 1979). Also see *North-South, A Program for Survival* (The Brandt Report) (London, Pan Books, 1980).

2. P.N. Giraud, *Geopolitique des Resources Minieres* (Paris, Economica, 1983).

3. J.S. Carman, *Obstacles to Mineral Development, A Pragmatic View* (New York, Pergamon, 1979).

and low prices.[4] Second, and independent of the first, the state-owned firms are much less flexible than private enterprises in adjusting capacity utilization to falling demand during economic recession. Especially where the industry has a heavy weight in the national economy, strong pressures will exist to maintain employment and foreign exchange income by operating the industry at full capacity, even at very low price levels.[5] The unwillingness to cut supply during recessions will result in lower average prices over the business cycle.

Those who hold these views usually believe that state enterprises' excessive capacity expansion and uneconomic operation at full capacity are made possible by concessional loans from their governments or from international agencies. Private mining firms do not have access to corresponding concessional capital. As a result, they may suffer financially from the lowered prices and could face severe difficulties in coping with competition on such uneven terms. If so, then in the long run, the emergence of the state-owned sector is seen as a threat to the very existence of the independent private mineral industry.[6] This purported threat leads to calls for policy intervention to restore the competitive balance. Such interventions might include restraints on the allocation of funds to international agencies, such as the World Bank or International Monetary Fund, that help finance mineral activities in developing countries, and protection or subsidies for the privately owned mines in industrialized importing countries.[7]

The above suppositions provide strong justification for a thorough and systematic study of the state mineral enterprise phenomenon to discover, as far as is possible, whether these worries are warranted and whether the suggested policy actions are justified.

A final reason for writing this book is the frequent disregard of state enterprises in recent scholarly works on worldwide mineral markets.[8] The authors of such works regularly presume that all market agents behave like private

4. R. Prain, *Copper, The Anatomy of an Industry* (London, Mining Journal Books Ltd., 1975).

5. *Metal Bulletin Monthly*, June 1983, p. 21; Rudolf Wolff & Co. Ltd., *Prospects for Copper*, December 1982; Phelps Dodge *Annual Report* 1982.

6. Chemical Bank, *Copper, the Next Twenty Years*, November 1981; Economist Intelligence Unit, *Raw Material Prices After the Recession*, April 1983; *Herald Tribune International*, October 15, 1983; *Mining Journal*, December 9, 1983; Metallgesellschaft, "Pressmeldungen über die Metallmärkte" (Frankfort am Main, January 1984).

7. *Mining Journal*, November 11, 1983.

8. See for instance D.L. McNicol, *Commodity Agreements and Price Stabilization* (London, Lexington, 1978); F.G. Adams and S.A. Klein, eds., *Stabilizing World Commodity Markets* (London, Lexington, 1978); and B.P. Bosworth and R.Z. Lawrence, *Commodity Prices and the New Inflation* (Washington, Brookings Institution, 1982).

profit maximizers. Their conclusions will have doubtful validity if, in fact, state mineral enterprises have behavioral characteristics that differ from that assumption. The present work provides some of the building blocks required to address such questions.

The international markets treated in the following chapters all have an oligopolistic supply structure. As table 1-1 shows, the capacity of the biggest firm in each mineral industry typically accounts for 10 percent or more of the Western World total. In this volume, "Western World" is defined as including all countries outside the centrally planned economies of Eastern Europe, Asia, and Cuba.

A certain degree of producer coordination has been common in many international mineral markets, both in market supply and, even more, in capacity expansion. Examples of such coordination include the prevalence of price setting by producers for a number of minerals (producer prices) and the substantial share of capacity expansion that has taken the form of joint ventures among many producers.

The entry of state-owned enterprises has affected the oligopolistic structure of mineral markets because the national takeovers have changed the degree of industrial concentration. The impact appears to have been both ways, however, reducing the degree of concentration in some markets and increasing it in others. The oligopolistic conduct in mineral markets has been weakened by the entry of new and inexperienced state enterprises that needed time to acquaint themselves with their competitors and develop oligopolistic relations with them. Having done so, the state-owned units have commonly displayed attitudes toward producer coordination that do not differ substantially from those of the private multinational mining firms. Hence, throughout this study, the working hypothesis is that the introduction of state ownership has not caused a permanent change in the oligopolistic structure or conduct in international mineral markets and that a degree of producer coordination prevails despite the altered ownership structure.

The resources available for carrying out the research that underlies this book did not permit more than three case studies. A number of factors contributed to the choice of PT Timah (tin) in Indonesia, Ferrominera (iron ore) in Venezuela, and ZCCM (copper) in Zambia for closer scrutiny. These cases illustrate a variety of minerals, geographical regions, and levels of economic development in the state-owned companies home countries. The cases also vary in length of period of state ownership. These companies predominately mine and process a single mineral, and state-ownership arrangements were

Table 1-1. Share of Western World Capacity for Leading Minerals Companies Ranked by Size
(percent)

Company rank	Aluminum		Copper		Iron ore mining	Lead refining	Nickel, Ferronickel	Tin smelting	Zinc reduction
	Mining	Smelting	Mining	Refining					
1	22.0	13.1	11.1	8.3	12.0	7.7	36.8	29.6	12.2
2	13.2	13.0	10.7	7.8	8.7	7.0	15.3	17.1	5.5
3	6.8	9.4	8.5	5.9	8.5	6.9	14.1	10.7	4.8
4	6.6	7.8	7.3	5.4	8.1	6.1	6.6	11.1	4.7
5	6.0	6.8	5.5	5.4	7.7	6.0	5.5	8.5	4.5
6	4.9	5.2	5.5	4.9	7.1	5.3	5.0	6.0	4.3
7	4.9	3.3	3.9	4.8	5.2	4.7	4.9	3.8	3.8
8	4.2	3.3	2.9	3.9	4.8	4.5	3.4	2.1	3.7
9	3.7	3.0	2.7	3.0	4.1	4.4	3.1	—	3.6
10	1.4	2.7	2.2	3.0	3.2	4.1	2.5	—	3.5
Total for companies listed	73.7	67.6	60.3	52.4	69.4	56.7	96.7	88.9	50.6

Note: Percentages shown include proportional shares of joint ventures.
Source: Mineral Processing in Developing Countries, U.N. Industrial Development Organization (UNIDO) 1980, U.N. Sales No. E.80.II.B.5.

reasonably clear-cut. The final choice was also influenced by my personal contacts and acquaintances within the mineral industry.

No far-reaching generalizations can be based on a sample of only three. An alternative set of cases—for instance, Bolivian tin, Brazilian iron ore, and Jamaican bauxite—might have produced a different set of conclusions. The one circumstance that may bias the case studies is that two of the three companies are located in OPEC countries. Both in Indonesia and Venezuela, public-sector financial conditions were very relaxed through most of the 1970s. This factor will be kept in view when the conclusions and generalizations from the case study sample are formulated in chapter 7.

The book consists of three parts. The first contains three chapters that present a general perspective on the issues treated. Chapter 2 explores the motivations for setting up state mineral enterprises, discusses alternative measures to quantify state ownership, and presents statistics to depict the growth and present significance of the state-owned sector in the production of different metal minerals. Chapter 3 is central to the entire study. It presents a survey of the general literature on state enterprise and derives what appear to be relevant and useful insights about the subset of publicly owned firms dealt with in the present study. These insights, combined with my own observations of the mineral industries over many years, are then analyzed to derive a set of hypotheses about the implications for such factors as costs, prices, and investment patterns in international mineral markets in which state-owned units from developing countries have come to account for substantial proportions of overall supply.

The second part of the book, comprising chapters 4, 5, and 6, consists of detailed case studies of the three state-owned mineral industries. The purpose of these studies is to provide some of the empirical evidence needed to test the hypotheses put forward in chapter 3.

The third part, chapter 7, pulls the different threads together, examines the extent to which the empirical findings contained in the case studies are in agreement with the hypotheses about market impact and brings out the major conclusions.

The findings and results of this study are tentative. As chapter 3 will show, at present very little established theory exists on state enterprise in general, and the empirical support for those theses that have been put forth is scattered and unorganized. This characterization of the state of the art certainly applies to the particular subject of state enterprises in the mineral industries. To my knowledge, work in this wide area is limited to the few contributions by Gillis, Labys, Mikesell, and Vernon.[9]

The dearth of earlier research makes the hypotheses formulated in this book highly preliminary. Although the three state corporate portraits provide useful qualitative insights about the validity of the hypotheses, they are too narrow a sample and contain an insufficient data base for definitive verification. In sum, then, the book presents a set of plausible hypotheses that have been derived from general empirical observation and economic analysis and that are supported by the case study material but that remain to be formally validated. Further research in this area is certainly warranted. Based on more complete evidence, such research could well lead to different findings and conclusions. Until this research has been completed, however, the present study is at least a step toward a better understanding of the consequences of the recently established state ownership positions for international mineral markets.

9. M. Gillis, "Allocative and X-efficiency in State-owned Mining Enterprises" (Cambridge, Mass., Harvard Institute for International Development, April 1979); M. Gillis and R.E. Beals, *Tax and Investment Policies for Hard Minerals, Public and Multinational Enterprises in Indonesia* (Cambridge, Mass., Ballinger Press, 1980); W. Labys, "The Role of State Trading in Mineral Commodity Markets: Copper, Tin, Bauxite and Iron Ore," Les Cahiers du CETAI, No. 79-06 (Montreal, 1979); R.F. Mikesell, *New Patterns of World Mineral Development* (London, British North American Committee, 1979); R. Vernon, "State-owned Enterprises in Latin American Exports," in W. Baer and M. Gillis, eds., *Export Diversification and the New Protectionism* (Champaign, Ill., University of Illinois Press, 1981); R. Vernon, "Uncertainty in the Resource Industries: The Special Role of State-owned Enterprises." Draft (Cambridge, Mass., Harvard University Center for International Affairs, January 1982); R. Vernon and B. Levy, "State-owned Enterprises in the World Economy, The Case of Iron Ore," in L. Jones et al., *Public Enterprises in Developing Countries* (Cambridge, Eng., Cambridge University Press, 1982).

2

Motivations for Establishing State Mineral Enterprises and Quantitative Assessment

The present chapter has two major purposes. The first is to explore the motivations for the growth and spread of state ownership in the mineral industries. The second is to establish the quantitative dimensions of state mineral enterprise.

The Nature of State Enterprise

"State enterprise" may mean different things to different people. The loose definition given below follows closely that of Gillis and identifies state enterprise by three key characteristics.[1] The first concerns the public aspects of the enterprises under study. The second and third distinguish these agents from others in the public sector by their enterprise features and the nature of their output.

The Ownership/Control Function

In a state enterprise, the government is the principal owner or exercises control over the broad policies of the enterprise, such as decisions on large investments or on additions of new lines of activity, and has power to appoint and dismiss management. The ownership function unambiguously distin-

1. M. Gillis, "The Role of State Enterprises in Economic Development," Discussion Paper 83 (Cambridge, Mass., Harvard Institute for International Development, February 1980).

guishes state enterprise from private enterprise. The control function also identifies as a state enterprise any operation in which the government does not have formal majority ownership but does exercise policy and managerial control. Such rights to control, implying a dilution of the private ownership rights, may follow from legislation or from negotiated agreements with the formal owners.

The Activity Function

A state enterprise engages in the production of goods and services for sale to the public or to other enterprises. This activity function distinguishes state enterprises from public undertakings, such as judicial and police systems, defense, and fire protection, which have far-reaching external effects that make these outputs public goods, unsuitable for direct sale to users.

The Revenue/Cost Function

This characteristic underlines the basically commercial nature of state enterprise. As a matter of policy, the sales revenues of the enterprise should bear a reasonably strong relation to cost. Thus, the postal services or a subsidized public transport system, which differentiate their charges according to the costs of services rendered, would be classified state enterprises. In contrast, a public hospital charging a flat fee, irrespective of the treatment requirements, would not qualify as a state enterprise.

State Enterprise in the Economics Literature

The particular state enterprises considered in the present study are firms that produce minerals and that were established in developing countries during a particular phase of these countries' histories. Before considering the forces that led to the emergence of state mineral enterprise in developing countries, it may be fruitful to summarize what the economics literature has to say about the motivations for and justifications of state entrepreneurship in general.

State ownership of factors of production is the rule in Socialist countries. No special motivations are needed to set up state enterprises in such economies. In market economies, however, private enterprise is presumed to satisfy the need for production and distribution of goods and services in an economically efficient way. In the process, private enterprise is also expected or required to meet a variety of noneconomic requirements sanctioned by general custom or law, for example, limitation of work hours, provision of old age pensions, assurance of safe working conditions, or, more recently, restriction of environmental pollution. Only if private enterprise fails to accomplish its task as efficient producer or if some special noneconomic goals must be satisfied in the course of the production process does motivation to set up state-owned enterprises arise.

The economics literature cites several general justifications for setting up state enterprise. These economic arguments do not address the real reasons political systems, in fact, establish such enterprises. They merely identify problems that are not automatically attended to by private enterprise. These arguments do not infer that founding a state enterprise affords the first-best solution to the problems nor that the state enterprise will actually resolve the problems at all.

Market Failure Motivations

Under certain circumstances, private enterprises in competitive markets will accomplish the optimum achievable level of welfare in society. Economists refer to this as the Pareto optimal condition.[2] When a Pareto optimum has been attained, the utility of any agent in the market cannot be increased without reducing the utility of some other agent. Market failure is considered a circumstance in which the market, when left to itself, fails to attain a Pareto optimum. In cases of market failure, public action might create conditions allowing improvements in welfare without compensating losses. The appropriate public action can take different forms, one of which is to set up a state enterprise.

Market failures can be both static and dynamic. The two major static failures that could justify the establishment of state enterprise stem from a noncompetitive market structure and from economic externalities.

UNCOMPETITIVE MARKET STRUCTURE. Lack of competition may exist among either sellers or buyers in a market. Economies of scale are a common cause of oligopolistic or monopolistic market conditions. Other causes of market concentration include appropriable managerial, organizational, technical, and natural resources—for example, an established marketing and dis-

2. For the necessary preconditions, see any standard microeconomic text such as P.R.G. Layard and A.A. Walters, *Microeconomic Theory* (New York, McGraw Hill, 1978).

tribution system, patents, or ownership of rich mineral reserves. The result of market power may be a restriction of production, with excessive output prices or depressed input prices. Another consequence could be internal inefficiency in production, resulting from restricted access to technological, managerial, and organizational inputs or from lack of competitive pressure in the output market.

The government can induce oligopolistic producers to behave competitively by entering the market and announcing that it will react to the private agents' output restrictions by expanding its own output, thereby driving prices to the competitive level. Or else a more efficient public enterprise can be set up to provide the private firms with a yardstick for feasible costs and prices.[3]

Alternatively, the government may decide to nationalize the entire sector and operate as a state enterprise to avoid the socially undesirable output restrictions and excessive price and profit levels reaped by private oligopoly or monopoly. Instances of such state intervention are found in utilities, telecommunication, some modes of transport, as well as in natural resource exploitation that derives high economic rents from rich intramarginal mineral deposits. If, however, the private monopoly is based on managerial or technical superiority internal to the firm and not available to the government, state enterprise may prove to be an inefficient tool for overcoming the detrimental effects of market failure.

ECONOMIC EXTERNALITIES. Private enterprise will not attain the socially desirable output level if substantial positive externalities exist in a given production activity and will exceed that level in the presence of negative externalities. Production under public ownership or control that takes account of costs and benefits external to the firm can rectify these shortcomings of private enterprise.[4]

Even if private enterprise attains Pareto efficiency in the static sense, dynamic market failure can occur if private firms are unable to realize potential future opportunities.

Perhaps the most important class of dynamic market failures results from excessive risk aversion and short-sightedness of private entrepreneurs, which prevent them from entering into new activities or ventures that take a long

3. F.M. Scherer, *Industrial Market Structure and Economic Performance* (Chicago, Rand McNally, 1980) pp. 486 and 487.

4. F.L. Pryor, "Public Ownership: Some Quantitative Dimensions," in W.G. Shepherd, ed., *Public Enterprise, Economic Analysis of Theory and Practice* (London, Lexington, 1976).

time to mature. Under these circumstances, state involvement in setting up infant industries is a time-honored practice. This involvement can take the form of direct ownership, tariff protection, subsidized loans, or price guarantees.

A special case of dynamic market failure has frequently occurred in recently independent countries. Under colonialism, private entrepreneurship may have been suppressed in size, international outlook, and technological sophistication. With independence, the establishment of new state enterprises has often been seen as the fastest, or even the only, way to establish large-scale or technically sophisticated activities or to diversify foreign trade relationships.

Political Motivations

Even in those market economies in which private enterprise seems to manage its tasks reasonably and therefore ideological or economic pressures for state-owned operations are not present, a number of cases can be singled out in which the state establishes an enterprise in response to particular political situations or a perceived need for this tool in the pursuit of specific political aims.

One such political condition has to do with national economic emancipation from colonial bonds following political independence. By far the most common circumstance under which developing countries have established state enterprises has been to take over foreign-owned operations having no obvious national private-sector claimants to the assets. Note that this phenomenon is not exclusive to the developing countries. Industrialized countries have on many occasions, especially after the great wars, confiscated foreign enterprises. In some cases, the confiscated assets have remained permanently in government ownership.

A political case for establishing state enterprise, both in industrialized and developing countries, is often made in cases in which an economic activity appears to be so important that decisions taken by private enterprise could compromise the independence and freedom of action of the political system. Common examples of industries nationalized for this reason include weapons, airlines, and postal services. Another related political motivation for establishing a state enterprise is a situation in which the government is the major or even the sole buyer of the industry's output. Weapons and certain health services are two examples. Here, too, the political rationale is to reduce dependence on private enterprise decisions,[5] but economic arguments based on advantages of vertical integration may further support the justification of state ownership.[6]

If for some reason an appropriate set of taxes and subsidies is not feasible for pursuing political and social goals like regional balance, promotion of the interests of suppressed racial groups, employment, or redistribution of income, state enterprise is often established to provide a second-best tool.[7] Such goals are related to broader than economic externalities and are not necessarily automatically achieved by profit-maximizing private enterprises in competitive markets. In some countries, state-owned companies are specifically entrusted with regional development tasks. This is true of the Instituto per la Reconstruzione Industriale (IRI) in Italy and the National Enterprise Board in the UK.[8] In many cases, regional policies are pursued by directing the state-owned railway system to impose below-cost charges for transport in sparsely populated areas. An important aspect of labor market policies in some countries has been to take over failing private firms to maintain employment and to continue operations with support from public subsidies.

State Mineral Enterprises in Developing Countries

The reasons for establishing state-owned enterprises, in the focus of the present study, tally reasonably with most of the motivations for public entrepreneurship in general. The special case among the mineral industries consists, in the main, of different motivations, caused partly by the nature of these industries. Also important, however, is the unique historical process of economic emancipation of the developing countries, which now account for a large proportion of state-owned mineral enterprises.

The mineral sector throughout the world appears to be a favorite area for government intervention in a variety of forms. The special interest of govern-

5. F.L. Pryor, "Public Ownership."

6. L. Jones and E.S. Mason, "The Role of Economic Factors in Determining the Size and Structure of the Public Enterprise Sector in Mixed Economy LDCs," in L. Jones et al., *Public Enterprises in Developing Countries* (Cambridge, Eng., Cambridge University Press, 1982).

7. I.B. Sheahan, "Public Enterprise in Developing Countries," *Public Enterprise*. Also see S. Holland, "State Entrepreneurship and State Intervention," in S. Holland, ed. *The State as Entrepreneur* (London, Widenfelt and Nicholson, 1972).

8. Swedish Government, Department of Industry, *State Business, Public Enterprise Experience in the EEC* (Stockholm, 1979).

ments in mineral activities has a number of explanations. First, the widespread belief that a country's mineral wealth is the nation's patrimony is used to sanction government intervention, for example, to prevent private interests from profiting by exploiting such assets or to assure that adequate supplies are left for future generations. Second, at least the mining stage is strictly tied to the mineral deposit and cannot move in response to rough treatment by the authorities; this affords the government considerable scope for intervention with impunity. Third, the general fiscal system is often an inefficient tool with which to appropriate mineral rents that regularly emerge from the exploitation of intramarginal deposits—therefore, governments often use special taxes or other means to intervene. And fourth, the extraction and processing of minerals are frequently regarded as strategically important because these activities assure domestic supply of key inputs into most manufacturing, including the defense industries. These characteristics of mineral production are regularly used to justify public intervention and control.

The second reason why minerals industries in developing countries are a special case has to do with the thorough economic emancipation that the Third World has experienced in all areas in recent decades. With improving public administrative, technical, economic, and managerial capabilities in the post-colonial period, ambitions to promote development through central control and direction of the economy expanded. The importance of the mineral sector in many countries and its predominantly foreign ownership, independence, and separation from the rest of the national economy made it one of the main targets for public policy initiatives.

These interventionist policies took a variety of forms. The rights to explore and exploit mineral deposits were curtailed and their terms stiffened. The fiscal dues were increased to absorb a significantly larger proportion of the surplus the mineral enterprises generated. The free movement of foreign exchange was restrained. Government control over the investment activity was extended to include approval and to impose local-purchase requirements. In many cases, the authorities intervened in production-level decisions and pricing matters, especially if the unit located in the developing country was part of a vertically integrated multinational chain. Targets were established for substituting nationals for foreign personnel in management positions.

The ultimate and most far-reaching intervention measure was for the government to become the owner of the industry. In most cases, the motivation to nationalize was founded in a view that other intervention measures were inadequate and that only direct, equity ownership could provide the government with effective control over this key industry, and hence over the entire economy. Effective control was seen as a necessary prerequisite for national development efforts.

The one-time nature of economic emancipation in developing countries, and its importance in motivating nationalizations in the mineral sector, suggest that the fast expansion of state ownership in developing countries' mineral industries, too, is a one-time phenomenon of limited duration.

Other general motivations for state enterprises have also had validity in the case of minerals in developing countries. This is true, for instance, of market failure resulting from uncompetitive conditions. Two distinctly different cases of lack of competition have been common. The importance of the mineral sector in the overall economy and its highly specialized needs often made it a dominant and sometimes sole buyer both of many domestic supplies and of skilled labor. Transfer of ownership into public hands was seen as a measure to prevent depressed input prices resulting from the buyers' market power or alternatively to let the government benefit from the monopsony profits.

The second case, which is less general, pertains to mineral monopoly. In contrast to nonintegrated mineral producers, who face fierce rivalry from other producers of the same commodity in the world market and from substitute products, the output and pricing decisions for mines in developing countries belonging to vertically integrated multinational resource companies are internal to these firms and may run counter to the interests of the nation. Such circumstances have provided a rationale for breaking up the vertical integration by putting the units in the developing countries under public national ownership.

State takeovers of mineral ventures were no exception from nationalizations in general in developing countries in that they almost invariably were directed at foreign-owned firms. The lack of entrepreneurial experience in the private sector, along with widespread socialist political leanings favoring a large and strong public sector, provided the motivation for public rather than private national ownership of foreign assets.

Large size and great significance to the national economy have particularly motivated nationalization of mineral ventures in developing countries. Industrialized economies are highly diversified; even in those heavily involved in mineral production—Australia and Canada, for example—manufacturing value-added is much greater than that generated by minerals. In distinction, many developing countries with rich mineral potential are heavily specialized, with minerals accounting for very high shares of GNP and exports; examples include Bolivia, Chile, Guinea, Papua New Guinea, Zaire, and Zambia.[9] Furthermore, in economic terms, a majority of developing countries are very small. For instance, in 1980, the thirteen nations of South America had a total GNP of $490 billion, while in the whole of Africa, with more than fifty nations, GNP did not exceed $350 billion. The GNPs of the medium-sized industrialized countries, such as Spain, Canada, and France, amounted to $200, $240, and $600 billion, respectively, in 1980.[10] Scale economies in the mineral industries make even individual large ventures weigh very importantly in the national economies of most developing countries. The case for public ownership in the mineral sector was especially strong because these countries' national investments, foreign exchange earnings, public revenues, and GNP were seen to stand and fall with their mineral activity.

Yet another of the general motivations for state ownership, of great importance for the case at hand, was the desire to use mineral enterprises as instruments in the pursuit of sociopolitical goals. The sizable rents mineral exploitation generated in many cases, along with the alleged insensitivity of private foreign owners to a variety of social needs, were seen to provide a strong justification for nationalization.

The same motivations that apply to state takeover of existing foreign-owned mineral enterprises in developing countries also motivate public ownership in the new projects that have been launched since the mid-1960s. In many new ventures, the assumption of complete ownership and full responsibility for management and control was not practicable either because the government found it financially burdensome or because the presence of the multinational mining firm was seen to be necessary to assure efficient operation. These factors have led to the emergence of a variety of hybrid arrangements, sometimes referred to as "new forms of investment," such as joint ventures, with the government holding only part of the equity; management contracts that entrust operational responsibility to an experienced foreign party; licensing agreements that assure continuous transfer of technology to the nationalized enterprise; or lease and production-sharing arrangements in which the foreign party exploits the government's mineral property for either a cash fee or a predetermined share of output. Lease and production-sharing arrangements have been common in oil and gas but not in metal minerals. Some of these hybrids, such as management contracts, have proved to be transient; others, such as joint ventures, appear more lasting.

9. G. Nankani, "Development Problems of Mineral Exporting Countries," Staff working paper no. 354 (Washington, World Bank, August 1979).

10. *World Bank Atlas* (Washington, World Bank, 1983).

Defining State Mineral Enterprise

Of the three broad components for defining state enterprises in general—(a) the ownership/control function, (b) the activity function, and (c) the revenue/cost function—the second and third can be applied without ambiguity to state mineral enterprises in developing countries. Without exception, such enterprises produce goods and services for sale to the public or to other enterprises, and their sales revenues do bear a reasonably strong relation to cost. The first component, ownership and control, is less clear.

The concept of control is crucial for defining state enterprise. If state and private enterprises differ systematically in behavior, the difference is that they are more likely to depend on control than on formal ownership positions. Although ownership can be passive, control implies active involvement.

The control concept raises some difficult definitional issues, however. First, how does one draw the line between that content and degree of state policy and managerial control that warrants classifying a firm as a state mineral enterprise and that which does not? Most developing country governments have become more economically interventionist in the past couple of decades, particularly with respect to large resource projects with a heavy weight in the national economy. Does this trend imply that all mineral enterprises in the most interventionist developing countries should be reclassified as state enterprises even if they continue to be wholly privately owned? Even if general government policies applied across the board were considered "control" for the purpose of identifying state enterprises, the implications of the general policy change would be to blur the distinctions in the behavioral characteristics of state-owned and privately owned enterprises.

A second issue is that the degree of control will be very hard to observe. The degree of formal control may differ from the extent to which it is exercised in practice. Even if the government has instituted far-reaching rules to assure its influence over the mineral activity, it may be unwilling or lack the ability to exercise that influence fully. In some cases, this situation would be easily noticeable—for instance, where most of the controlling powers have been handed over to a private corporation through a management contract. In most cases, determining the extent of government control would require in-depth studies of individual mineral enterprises.

For these reasons, a quantitative assessment of state enterprise, primarily based on the concept of control, would be quite difficult. Therefore, most such assessments are based on the extent of government equity ownership rather than of control. The presumption is that ownership and control are closely correlated.

The definition of state enterprise on the basis of the government's equity holding, although much simpler than that based on the concept of control, is not without complications. Owners are not always easily identified. For instance, government equity may be held by institutions not unambiguously recognized as part of the state. Another problem arises in the common case in which the mineral enterprise is jointly owned by private and public interests. One way to tackle this ambiguity is to define as state enterprise only those firms in which government holds more than half the total equity. Another is to allocate capital assets, capacity, employment, or production of each enterprise to the state and private sectors according to the equity shares held by each. Each definition will produce a different measure of the size of the state mineral sector.

Inclusion of only majority-owned firms in the state enterprise category may seem too restrictive. The division of enterprises into state and private portions, though convenient for the purpose of aggregate quantitative measurement, impedes clear-cut identification of individual enterprises as either state or private.

Yet a third way to use equity ownership for enterprise classification is to include all enterprises with "significant" government equity interest in the state enterprise category on the assumption that even a minority holding of equity, along with the government's powers to design and implement general economic policies, assures it of a high degree of influence and control over the enterprise. Even if the government does not use its powers overtly, its ownership involvement is sometimes seen to be a signal of interest and intent to intervene should the need arise. The response of the private equity holders in such cases may be to accommodate the government in order to avoid confrontations they know they could lose. The full influence of the government may be completely covert in such cases. An equity holding of 5 to 10 percent or more by the government is often considered "significant" in this context.

Classifying state enterprise by significant government interest in it also raises problems. Although public authorities may hold a significant equity position, the degree of government involvement in a mineral enterprise may vary from complete to none at all, depending on the government's attitudes and competence. The degree of government involvement also depends crucially on whether the private share of equity is widely dispersed among the general public at home or abroad or held by a powerful and experienced multinational mining corporation.

The presumption of a close correlation between public ownership and exercise of control is not always valid. For instance, although government holds some 20 percent of the equity in the Bougainville copper mine in Papua New Guinea, more than one-third of the equity in the uranium mines of Niger, and about 50 percent of the Lamco iron ore venture in Liberia, none of these authorities have overtly exercised their ownership powers to influence the policies and operations of the three mineral firms. In contrast, the Jamaican government has intervened in bauxite enterprises with regard, for instance, to price, investment, and forward processing policies, even with the enterprises in which it holds less than 10 percent of the equity. Of course, the difference in government behavior could also be due to the fact that the companies' foreign owners have chosen policies of accommodation in the first cases and of confrontation in the second.

The appropriate criterion to distinguish between state and private enterprise from the point of view of this study ideally would be the extent of government control. The ambiguities of the government control concept and the immense problems of measuring control make this criterion less suitable than government equity for quantitative assessments of state enterprise in the mineral sector. Although the measurement problem becomes much easier when government equity is used as the criterion to distinguish state from private enterprise, alternative approaches can be used for quantitative assessments, but none of them is free from ambiguities. The size of the state enterprise universe will vary depending on whether it is defined to include (a) all the units in which government has a "significant" equity position, (b) the proportions of individual enterprises corresponding to the government equity share in each case, (c) those enterprises in which government holds a majority, or (d) only the firms that are 100 percent owned by government.

Given the ambiguities of the control concept, the following section, which reports on attempts to measure the prevalence of state enterprise in individual mineral industries, relies entirely on equity ownership as the measuring rod. The share of government in Western World capacity in aluminum, copper, and iron ore is measured in terms of the first three definitions in the preceding paragraph. For several other mineral industries, the figures provided refer in the main to the second—that is, the capacity that is proportional to the government equity share in each individual enterprise.

Some Quantitative Assessments

Most of the figures in this section provide a snapshot view of the significance of state enterprise in a group of important mineral markets from a recent year, although some discussion of the growth of state enterprise over time is also included. No large-scale state divestments have been recorded during the past few decades in the industries surveyed, so growth has been positive throughout. Only the Western World capacity or production, as the case may be, is considered in the assessments.

Comprehensive figures have been compiled for the three most important mineral industries—aluminum, copper, and iron ore. The reader interested in the detailed results is referred to the appendix to this chapter. Table 2-1 presents a summary of these compilations.

Table 2-1. Government Ownership Positions in Three Major Mineral Industries, 1981

	Aluminum (actual weight)[a]			Copper (metal content)			Iron ore
	Mining	Refining	Smelting	Mining	Smelting	Refining	(actual weight)[b]
Western World							
Total capacity (000 tons)	92,500	30,790	14,040	7,820	8,780	9,120	543,000
Percent of capacity with significant govt. ownership[c]	45.3	24.2	22.8	40.5	30.6	25.9	55.6
Percent of equity held by government	27.8	15.1	18.5	32.4	26.1	21.6	40.0
Percent of capacity with majority government ownership	23.2	13.2	19.4	34.7	29.7	24.3	n.a.
Developing countries							
Total capacity (000 tons)	54,000	6,530	2,190	4,120	3,340	2,580	216,900
Percent of capacity with significant govt. ownership[c]	71.1	54.9	60.3	73.0	75.5	82.6	94.7
Percent of equity held by government	41.1	21.1	44.7	57.8	64.0	67.6	61.8
Percent of capacity with majority government ownership	33.3	12.3	45.6	62.0	72.9	77.0	n.a.

Source: Computed from tables 2.A-1 through 2.A-7.

[a] Figures for aluminum are for 1980.

[b] Figures for iron ore are based on production and not on capacity.

[c] Defined as 5 percent and upwards of total equity.

Bauxite/Alumina/Aluminum

The bauxite figures in table 2-1 demonstrate the large differences in the measures of the state-owned share, depending on how state enterprise is defined. No less than 45 percent of Western World capacity has "significant" government ownership, but the figure drops to 28 percent if state ownership is measured in proportion to government equity holdings and to 23 percent if only capacity with majority government ownership is included.

State ownership in bauxite mining is heavily concentrated in the Third World. With the exception of Yugoslavia (capacity 3.5 million tons), all the companies with government equity involvement, listed in appendix table 2.A-1, are located in the Third World. On all three definitions, state ownership is more pervasive in developing countries than in the Western World.

Most of the state ownership positions in bauxite were established during the 1970s. Two forces have been involved. First, the state-owned sector has been expanded through total or partial government takeovers. For example, a complete nationalization of bauxite production in Guyana occurred in 1971, while the government takeover of majority shares in Kaiser's and Reynolds' bauxite operations in Jamaica dates to the mid-1970s.[11] As late as 1970, the industry was fully privately owned in both countries. The second force explaining the growing importance of state enterprise in bauxite is that in several of the countries with state-owned enterprises the industry has expanded rapidly. This is particularly true in Guinea, where capacity increased about sixfold, to 14.5 million tons, between 1972 and 1980. In Brazil, the Rio Norte project in Trombetas started production only in 1979 with a capacity of 3.4 million tons; doubling of capacity, planned by 1983, has been postponed due to depressed market conditions.[12]

In both alumina refining and aluminum smelting, the government shares are substantially lower than in bauxite. Also, although state ownership of bauxite capacity is by and large a developing country phenomenon, one-half or more of the state-owned capacity in alumina and aluminum is located in the industrialized world.

The relatively complex technology in alumina refining as compared to bauxite mining and aluminum smelting has permitted the aluminum multina-

11. R. Mikesell, *New Patterns of World Mineral Development*, British North American Committee, 1979.

12. *Mining Annual Review*, 1983.

tionals to keep a firmer grip on this stage of production.[13] The fear of ruptured supply of technology and ensuing losses of efficiency has discouraged developing countries from takeover actions. An additional important reason for the high multinational ownership share at the refining stage of production is the very low contribution of alumina refining to net foreign exchange earnings and to gross domestic product in comparison with bauxite mining and alumina smelting.[14]

Western Europe has substantial state ownership of aluminum smelting. Some of the state involvements in this region, for example, in the West German Vereinigte Aluminium Werke (VAW), were established in the reconstruction period just after the second world war. In Norway, the government became the owner of the Aadal and Sundal Verk (ASV) plants in the late 1940s as a result of confiscation of German property. In Italy, on the other hand, state ownership emerged only in the early 1970s as a result of industrial reconstruction.[15]

Most of the state-owned aluminum smelting in developing countries was established in the 1970s. In Venezuela, the large-scale smelting capacity with heavy government participation (400,000 tons) dates back to 1968 for the first smelter (Alcasa) and to 1978 for the second one (Venalum). The sharply increased energy prices in the mid-1970s provided strong inducements to locate new aluminum smelters in the Third World close to sources of cheap electrical power. Many of the new smelters have been established under government ownership. This is true of the installations in Bahrain, Dubai, Egypt, Iran, Turkey, and also in South Africa (total capacity 620,000 tons). None of these countries had any smelting capacity in 1970.

The appendix tables do not reflect a further boost to state ownership in the bauxite/alumina/aluminum industry that occurred in 1981. This was the nationalization of Pechiney-Ugine Kuhlman (PUK) in France, one of the six leading multinational companies in the industry. In 1977, this company owned 4.9 percent of total bauxite capacity, 8.6 percent of alumina capacity and 6.8 percent of aluminum capacity in the Western World. These figures include PUK's proportionate shares of capacity in joint ventures.[16]

13. R.G. Adams, "Structural Change in the World Aluminum Industry." Presentation to the Chase European Aluminum Seminar (Zurich, June 16, 1981).

14. H. Hashimoto, "Bauxite Processing in Developing Countries" (Washington, World Bank, January 1982).

15. Interview with the management of Gränges Aluminum, Sweden, March 1983.

16. *Mineral Processing in Developing Countries*, Sales No. E.80.II.B.5 (New York, United Nations, 1980).

Copper

Table 2-1 reveals that the differences among the three measures of state ownership are much smaller in the case of copper than in aluminum. The state-owned sector accounts for roughly one-third of copper mining and about one-quarter of copper refining, regardless of which measure is used.[17]

Like bauxite, but in contrast to alumina refining and aluminum smelting, state ownership at all stages of copper production is almost exclusively a developing country phenomenon. Finland and Yugoslavia are the only countries outside the developing country group with significant copper production under state ownership. The dominance of developing countries in the state-owned copper sector, in conjunction with the lesser importance of Third World government participation at the consecutive stages of copper processing, explain the slightly declining share of government ownership at the smelting and refining stages.

The dynamic expansion of the state-owned copper sector during past decades is easy to document:

> At the beginning of the [1960s] copper production in which government held any sort of interest did not amount to more than 100,000 tons per year, or 2.5 percent of capacity in the so-called 'free world.' By 1970, this total had risen to some 2.25 million tons, or about 43 percent of capacity. More than a quarter of the world's copper was being produced by mines totally owned by government, 12 percent by companies in which the State had a majority interest, and 5 percent by companies in which government had minority interests.[18]

Major features in this change, chronologically arranged, include[19]

a. the complete nationalization in 1967 of Gecamines in Zaire;
b. the 51 percent government takeover in 1969 of Zambian capacity and of the large mines (Codelco) in Chile;
c. the complete nationalization in 1971 of the large mines (Codelco) in Chile;
d. the nationalization in 1974 of Cerro de Pasco in Peru, subsequently renamed Centromin;

17. The Yugoslavian copper industry has been classified as state owned, though, strictly speaking, it is owned by its employees.

18. R. Prain, *Copper, the Anatomy of an Industry* (London, Mining Journal Books Ltd., 1975).

19. Based on M. Radetzki, "Copper Dependent Development," UNCTAD/LDC/11, 12 June 1980; and CIPEC information.

Table 2-2. Principal State-owned Ventures Exporting Iron Ore, 1978

Country	Name of establishment	Date established as state enterprise	Iron ore production, 1978 (000 tons, actual weight)
Brazil	Companhia Vale do Rio Doce (CVRD)	1942	50.574
Sweden	Luossavaara-Kiirunavaara AB (LKAB)	1907/1957[a]	23.967
South Africa	South African Iron & Steel Industrial Corporation (ISCOR)	1928	19.796
Liberia	Lamco Joint Venture	1960[b]	10.572
	Bong Mining Company	1963[b]	7.387
India	National Mineral Development Corporation (NMDC)	1958	6.909[c]
Venezuela	C.V.G. Ferrominera Orinoco S.A.	1974	12.956
Chile	Compañía de Acero del Pacifico S.A. (CAP)	1971	6.935
Mauritania	Société Nationale Industrielle et Minière (SNIM)	1974	6.336
Peru	Empresa Minera del Peru	1975	4.854

Source: Vernon and Levy, "State-owned Enterprises."

[a] The Swedish government held 50 percent of the company's shares in 1907; in 1957, the government took over the company entirely.

[b] The Liberian government has not attempted to exercise any control over these projects, which are effectively controlled by foreign partners.

[c] Along with the export operations listed here, iron ore mines captive to state-owned steel companies mine about 12 million tons of ore annually.

e. the startup of production in 1977 of Cerro Verde (Mineroperu), a new state-owned mine in Peru;

f. the increase in 1979 of the government equity share from 51 percent to 60 percent in Zambia; and

g. the startup of production in 1980 in the La Caridad mine (Mexicana del Cobre) in Mexico, in which the government holds 44 percent of total equity.

Iron Ore

The data on iron ore industry state ownership are less detailed than in the case of aluminum and copper. Appendix table A.2-7 presents the figures on state

Table 2-3. Production of Tin Concentrates, 1950 and 1980
(000 tons of tin metal)

Country	1950	1980
Malaysia	58.5	61.4
Thailand	10.5	33.7
Indonesia	32.6	32.5
Bolivia	31.7	27.5
Total, 4 countries	133.3	155.1
Total, Western World	164.5	199.8

Source: Metal Statistics, annual, various issues (Frankfort an Main, Metallgesellschaft).

ownership in the Western World. Only countries with government ownership have been listed. The percentages pertain to the government share of national production in 1981 and do not indicate the government share of equity in individual enterprises. The data therefore do not permit a quantification of production accounted for by units in which governments hold a majority of the equity.

State-owned production is 40 percent of the Western World total, if production is apportioned to the governments in accordance with their equity holdings. This share is considerably higher than at any stage of aluminum or copper production. The state-owned share in iron ore, measured in the same way, amounts to 62 percent for developing countries, about the same as in copper smelting and refining.

The growth of importance of state-owned enterprise in iron ore mining over time is illustrated by table 2-2, which is reproduced from a recent paper by Vernon and Levy.[20] The focus of this table is on the large and export-oriented state-owned firms. The Liberian operations have been 50 percent owned by the government since their establishment in the early 1960s. The Latin American and Mauretanian companies listed were transformed into state enterprises through nationalizations in the 1970s. The Brazilian Companhia Vale do Rio Doce has strongly expanded its share of world output by more than doubling its capacity in the course of the past decade.

Tin

Developments in individual tin-producing countries indicate that state participation in this industry is important. Table 2-3 reveals that four countries have

20. R. Vernon and B. Levy, "State-owned enterprises in the world economy, the case of iron ore," in L. Jones et al., *Public Enterprises in Developing Countries* (Cambridge, Eng., Cambridge University Press, 1982).

long dominated world production of tin concentrates. These countries' share of Western World output was 81.0 percent in 1950 and 77.6 percent in 1980.

In 1950, virtually all production capacity in three of the four countries was in private hands. The exception was Indonesia, where the Dutch colonial government was a majority owner of the tin industry. Important structural changes have occurred since that time. First, the three major tin mining groups in Bolivia were nationalized in 1952. Comibol, the state tin company, has since then accounted for between two-thirds and three-quarters of Bolivia's tin output.[21] In Indonesia, the entire industry was nationalized in the 1950s, and even though some private enterprises reemerged in the 1970s, the state tin corporation, PT Timah, accounted for 80 percent of the national tin output in 1980.[22] In Malaysia, the government took far-reaching steps during the 1970s to acquire important equity positions in the nation's tin enterprises. According to Thoburn, the Malaysia Mining Corporation, in which the government has a 71 percent ownership position, controls more than 25 percent of the country's tin output.[23] Thoburn also notes that the governments of Nigeria and Zaire have recently acquired large equity stakes in the formerly foreign-owned tin mines.[24] These two countries' share of Western World output in 1980 amounted to about 3 percent.

These figures suggest that state ownership in the Western World tin industry increased from almost nil in 1950 to at least 30 percent in 1980. Four-fifths of Western World tin capacity is in developing countries, and the current state ownership share in these areas would exceed 35 percent.

Other Minerals

The four minerals treated above represent two-thirds or more of the value of all metal minerals in the world economy. Although this survey is not exhaustive, some additional metal mineral industries are also subjects of state ownership.

COBALT. State-owned firms account for a very high share of cobalt output. Western World production in 1980 amounted to 26,600 tons.[25] The four coun-

21. J. Thoburn, "Policies for Tin Exporters," *Resources Policy*, June 1981; and J. Thoburn, *Multinationals, Mining and Development, A Study of the Tin Industry* (Brookfield, Vt., Gower Publishing Co., 1981).

22. See chapter 4.

23. J. Thoburn, *Multinationals, Mining and Development*, 1981.

24. J. Thoburn, "Policies for Tin Exporters," June 1981.

25. U.S. Bureau of Mines, *Minerals Yearbook*, 1981.

Table 2-4. Government Ownership in Cobalt Production, Selected Countries, 1980

Country	Output (000 tons)	Government equity (percent share)
Zaire	15,500	100
Zambia	4,500	60
Finland	1,040	81
Morocco	750	100

Sources: U.S. Bureau of Mines, *Mineral Industries of Africa*, 1976; U.S. Bureau of Mines, *Minerals Yearbook*, 1981; Outokumpu *Annual Report*, 1981.

tries listed in table 2-4, in which the state is the majority owner of companies producing the metal, accounted for 82 percent of the total. The state-owned share was as high as 61 percent, even when only the firms that are 100 percent state owned are included. The exceptionally high state share follows from the unique position of Zaire, which accounts for more than half of Western World output, all through a single company, Gecamines, which was fully nationalized in 1967.

LEAD AND ZINC. State enterprise has been assessed to account for only 20 to 25 percent of overall equity in lead and zinc enterprises.[26] This relatively low state involvement is partly explained by the fact that a very large share of Western World capacity is located in industrialized countries.

NICKEL. Private enterprise produces about 85 percent of Western World nickel output. No nationalization has occurred in this industry since the one in Cuba in 1960. About half of the Third World's nickel output comes from the French territory New Caledonia, where there is some French government participation. A government enterprise in Indonesia, Aneka Tambang, accounts for about 5 percent of total Western World output.[27] Also, governments have taken equity positions in a few smaller ventures in developing countries that went into production in the late 1970s.

CHROMIUM. The major state-owned chromium-producing units are found in Finland, India, Malagasy, and Turkey. Together, they account for approxi-

26. S. Harris, "The Commodities Problem and the International Economic Order: What Rules of What Game?," in M.P. Oppenheimer, ed., *Issues in International Economics* (London, Oriel, 1980).

27. R. Mikesell, *New Patterns of World Mineral Development*, British North American Committee, 1979.

mately 17 percent of Western World output. About 30 percent of developing countries' capacity is in government hands.[28]

MOLYBDENUM. The state-owned Codelco in Chile accounts for 14 percent of Western World and 90 percent of developing countries' output of molybdenum.[29]

Prospects

Despite their limitations, the data in this section clearly point to at least three general conclusions. First, despite large variations among individual minerals, the amount of state equity ownership in Western World mineral industries is very substantial. Second, it appears that state ownership is most prevalent in the developing countries; in several minerals, state enterprises dominate Third World capacity. And third, most of the state-owned positions have been established since 1950, with a strong spurt in the first half of the 1970s, when nationalization seemed to become contagious.

State enterprise, measured as capacity proportional to government equity holdings, has evolved from relative insignificance in the early 1950s to include in the early 1980s roughly one-third of the Western World mineral industry and about half of the mineral production capacity located in the developing countries. This important position has been attained primarily through a combination of nationalizations of existing privately owned firms, growth in capacity of existing state-controlled companies, and substantial government participation in joint ventures with mining multinationals in the development of new capacity.

Several factors suggest that the fast-growth phase of state-ownership in Western World mineral industries may now have come to an end. Political independence and the urge for economic emancipation in the Third World, which was a major force behind the many nationalizations in past decades, may have run out of force in the early 1980s. The major adjustments to rectify unfavorable colonial conditions had been completed, and the newly independent governments had already absorbed the most conspicuous foreign ownership positions. Socialist leanings in developing countries appear less in vogue now than in the 1950s and 1960s, and pragmatism has taken precedence over

28. M. Radetzki, "Strategic Metal Exports from Developing Countries: Policy Options to Increase National Benefit." Study for United Nations Conference on Trade and Development (Geneva, United Nations, December 1983).

29. P. Crowson, *Minerals Handbook 1982–83* (New York, Macmillan, 1982).

ideology.[30] Relationships between governments in developing countries and multinational mining firms have improved, and amicable joint-venture arrangements for the development of new projects have become common.[31] The emergence and increasing sophistication of private entrepreneurship in developing countries suggests a greater role for private national ownership in minerals, at least in future expansions of this sector.

In the industrialized market economies, in the early 1980s, the governments show no signs of extending their ownership positions in the mineral industries. Neither industrialized nor developing countries, however, show tendencies towards dismantling existing public equity involvements in these industries.

The tentative conclusion is, then, that the period of fast growth of state ownership may now be over and that in the decade to come an increasingly mature group of state-owned enterprises will be supplying a substantial, but relatively static, proportion of Western World mineral supply.

Appendix 2.A
State Ownership in Three Major Mineral Industries

Tables 2.A-1 through 2.A-6 list production capacities with significant government ownership in the aluminum and copper industry. Significant ownership has been defined as an equity holding of at least 5 percent. The tables indicate the percentage of equity held by government and the share of capacity that is proportional to government equity holding. The production capacity in which government equity exceeds 50 percent is also specified.

Table 2.A-7 shows similar information for iron ore, but the data are based on production rather than capacity. Also, because the information is by country and not by company, the final column, showing the capacity in which government holds a majority of the equity, is omitted.

30. UN Center on Transnational Corporations, "Transnational Corporations in World Development," Sales No. E.83.II.A.14 (New York, United Nations, 1983).

31. *Ibid.*

Table 2.A-1. Western World Bauxite Capacity with Significant Government Ownership, 1980

Country	Company	Total capacity (000 tons)	Government equity share (%)	Capacity proportional to government equity holding (000 tons)	Capacity of companies with government equity above 50% (000 tons)
Africa					
Ghana	British Aluminum Co.	300	55	170	300
Guinea	Guinea Bauxite Co.	9,000	49	4,410	—
	Friguia	3,000	49	1,470	—
	Kindia Bauxite Office	2,500	100	2,500	2,500
Asia					
India	Bharat Aluminum Co.	400	100	400	400
	India Aluminum Co.	500	45	230	—
	Madras Aluminum Co.	100	100	100	100
Indonesia	Aneka Tambang	1,800	100	1,800	1,800
Latin & Central America					
Brazil	Companhia Brasilieva do Alumino (CBA)	500	20	100	—
	Mineracao Rio Norte	3,400	41	1,400	—
Guyana	Guymine	5,000	100	5,000	5,000
Jamaica	Jamalcan	2,700	7	190	—
	Jamalco	1,270	6	180	—
	Kaiser Bauxite Co.	4,200	51	2,140	4,200
	Reynolds Bauxite Co.	3,100	51	1,580	3,100
Europe					
Turkey		600	100	600	600
Yugoslavia		3,500	100	3,500	3,500
Total		41,870		25,670	21,500
Total in developing countries		38,370		22,170	18,000

Note: Overall Western World capacity is 92.5 million tons, of which 54 million is in developing countries. *Source:* United Nations Conference on Trade and Development (UNCTAD), *Processing and Marketing of Bauxite/Alumina/Aluminum: Areas for International Cooperation*, TD/B/C.1/PSC/19, United Nations, August 1981.

Table 2.A-2. Western World Alumina Refining Capacity with Significant Government Ownership, 1980

Country	Company	Total capacity (000 tons)	Government equity share (%)	Capacity proportional to government equity holding (000 tons)	Capacity of companies with government equity above 50% (000 tons)
Africa					
Guinea	Friguia	700	49	343	—
Asia					
India	Bharat Aluminum Co.	200	100	200	200
	Indian Aluminum Co.	236	45	106	—
	Madras Aluminum Co.	50	73	37	50
Latin & Central America					
Brazil	CBA	200	20	40	—
Guyana	Guymine	350	100	350	350
Jamaica	Jamalcan	1,100	7	77	—
	Jamalco	550	6	33	—
Europe					
FRG	VAW	600	50	300	—
	VAW	640	100	640	640
Italy	Euroallumina, Unit 1	720	61	439	720
	Euroallumina, Unit 2	200	94	188	200
Turkey	Seydisehir	200	100	200	200
Yugoslavia		1,700	100	1,700	1,700
Total		7,446		4,653	4,060
Total in developing countries		3,586		1,376	800

Note: Overall Western World capacity is 30.8 million tons, of which 6.5 million is in developing countries.

Source: UNCTAD, *Processing and Marketing of Bauxite/Alumina/Aluminum*.

Table 2.A-3. Western World Aluminum Smelting Capacity with Significant Government Ownership, 1980

Country	Company	Total capacity (000 tons)	Government equity share (%)	Capacity proportional to government equity holding (000 tons)	Capacity of companies with government equity above 50% (000 tons)
Africa					
Egypt	Egyptalum	100	100	100	100
South Africa	Alusaf	88	66	58	88
Asia					
Bahrain	Alba	125	78	98	125
Dubai	Dubal	135	80	108	135
India	Balco	75	100	75	75
	Indalco	120	45	54	—
	Nalco	25	73	18	25
Iran	Iralco	50	94	47	50
Latin & Central America					
Brazil	CBA	82	20	16	—
Venezuela	Alcasa	120	50	60	—
	Venalum	280	80	224	280
Europe					
Austria	Raushofen	80	100	80	80
FRG	VAW	340	100	340	340
Italy	Societa Aluminio Veneto (SAVA)	60	50	30	—
	Aluminio Sarda (ALSAR)	125	100	125	125
	ALUMETAL	100	100	100	100
Norway	ASV	335	100	335	335
	Norsk Hydro	120	75	90	120
	Det Norske Nitrit Selskab (DNN)	60	100	60	60
Spain	Endasa	126	58	73	126
	Alugasa	92	26	24	—
	Al Espanol	180	57	103	180
Turkey	Seydisehir	120	100	120	120
Yugoslavia		260	100	260	260
Total		3,198		2,598	2,724
Total in developing countries		1,320		978	998

Note: Overall Western World capacity is 14.0 million tons, of which 2.2 million is in developing countries.
Source: UNCTAD, *Processing and Marketing of Bauxite/Alumina/Aluminum.*

Table 2.A-4. Western World Copper Mining Capacity with Significant Government Ownership, 1981

Country	Company	Total capacity (000 tons)	Government equity share (%)	Capacity proportional to government equity holding (000 tons)	Capacity of companies with government equity above 50% (000 tons)
Africa					
Botswana	Selebi Pikwe	17	15	3	—
Morocco	Various	8	100	8	8
Zaire	Gecamines	662	100	662	662
	Sodimiza	40	15	6	—
Zambia	Zambia Consolidated Copper Mines (ZCCM)	704	60	422	704
Asia					
India	Hindustan Copper	35	100	35	35
Malaysia	Mamut	28	49	14	—
Latin America					
Bolivia	Comibol	7	100	7	7
Brazil	Brasiliana de Cobre	30	100	30	30
	Min Sul Vicosa	4	51	2	4
Chile	Codelco	890	100	890	890
	Enami	25	100	25	25
Mexico	Macocosac	11	100	11	11
	Mexicana del Cobre	180	44	79	—
	Mina de Cananea	65	52	34	65
Peru	Centromin	34	100	34	34
	Mineroperu	33	100	33	33
Oceania					
Papua New Guinea	Bougainville	165	20	33	—
Europe					
Finland	Outokumpu oy	38	81	31	38
Israel	Timna	12	100	12	12
Spain	Apisa	14	100	14	14
Turkey	Black Sea	23	29	7	—
	Etibank	19	100	19	19
Yugoslavia	RTB Bor, Kapaonik, RB Bucim	100	100	100	100
Other countries[a]		26	100	26	26
Total		3,170		2,537	2,717
Total in developing countries		3,006		2,380	2,553

Note: Overall Western World capacity is 7.8 million tons, of which 4.1 million is in developing countries.

Source: Personal communication with Conseil Intergouvernemental des Pays Exportateurs de Cuivre (CIPEC).

[a] These countries are all developing countries.

Table 2.A-5. Western World Copper Smelting Capacity with Significant Government Ownership, 1981

Country	Company	Total capacity (000 tons)	Government equity share (%)	Capacity proportional to government equity holding (000 tons)	Capacity of companies with government equity above 50% (000 tons)
Africa					
Zaire	Gecamines	469	100	469	469
Zambia	ZCCM	816	61	494	816
Asia					
India	Hindustan Copper	48	100	48	48
Latin America					
Chile	Codelco	770	100	770	770
	Enami	190	100	190	190
Mexico	Mina de Cananea	85	26	22	—
	Various	3	100	3	3
Peru	Centromin	58	100	58	58
	Mineroperu	33	100	33	33
Europe					
Finland	Outokumpu oy	60	81	49	60
Portugal		7	100	7	7
Turkey	Black Sea	20	100	20	20
	Etibank	20	100	20	20
	Etibank Murgul	12	100	12	12
Yugoslavia	RTB Bor	100	100	100	100
Total		2,691		2,295	2,606
Total in developing countries		2,524		2,139	2,439

Note: Overall Western World capacity is 8.8 million tons, of which 3.3 million is in developing countries.

Source: Personal communication with Conseil Intergouvernemental des Pays Exportateurs de Cuivre (CIPEC).

Table 2.A-6. Western World Copper Refining Capacity with Significant Government Ownership, 1981

Country	Company	Total capacity (000 tons)	Government equity share (%)	Capacity proportional to government equity holding (000 tons)	Capacity of companies with government equity above 50% (000 tons)
Africa					
Zaire	Gecamines	151	100	151	151
Zambia	ZCCM	776	61	473	776
Asia					
India	Hindustan Copper	40	100	40	40
Latin America					
Chile	Codelco	620	100	620	620
	Enami	160	100	160	160
Mexico	Colere de Mexico	130	48	63	—
	Mina de Cananea	15	26	4	—
Peru	Centromin	55	100	55	55
	Mineroperu	183	100	183	183
Europe					
Finland	Outokumpu oy	60	81	49	60
Yugoslavia	Rudarsko Bor	175	100	175	175
Total		2,365		1,967	2,220
Total in developing countries		2,130		1,743	1,985

Note: Overall Western World capacity is 9.1 million tons, of which 2.6 million is in developing countries.

Source: Personal communication with Conseil Intergouvernemental des Pays Exportateurs de Cuivre (CIPEC).

Table 2.A-7. Western World Iron Ore Production with Significant Government Ownership, 1981

Country	Production (million tons, actual weight)	Government ownership share of national production (%)	Production proportional to government ownership share (million tons, actual weight)
Africa			
Algeria	3.0	100	3.0
Liberia	18.5	50	9.3
Mauretania	8.0	100	8.0
S. Africa	25.5	90	23.0
Asia			
India	40.0	50	20.0
Latin America			
Brazil	105.2	60	63.1
Chile	8.7	100	8.7
Peru	5.6	100	5.6
Venezuela	14.0	100	14.0
Europe			
France	22.3	85	19.0
Norway	4.1	85	3.5
Austria	3.1	100	3.1
Sweden	22.3	100	22.3
Spain	8.6	40	3.4
Turkey	4.9	100	2.1
Yugoslavia	4.9	100	4.9
Others with state ownership	5.9	75	4.4
Total	301.8		217.4
Total in developing countries	205.4		134.1

Note: Overall Western World production is 543.0 million tons, of which 216.9 million tons is in developing countries.

Source: Personal communication with Malmexport AB, Stockholm.

3

Characteristics and Market Impact of State Mineral Enterprises

Given the importance and recency of state mineral ownership, the next question is how does an important state enterprise presence affect the international mineral markets? This chapter sets forth some hypotheses concerning this question. These hypotheses are derived mainly from study of the literature and from close observation of the operations of the mineral industry in practice.

As background, a review of the general state enterprise literature is used to identify the special features that characterize state-owned firms. Little has been written specifically on state mineral enterprises in developing countries, so the broader literature is essential to formulating hypotheses about the importance of such agents in the marketplace. A central part of this chapter is devoted to delineating the characteristics and behavioral patterns of state-owned mineral firms in developing countries. The final section presents the hypotheses formulated on the basis of this analysis.

State Enterprise in the Economics Literature

The dominant approach in economic studies of state enterprise is to begin with the microeconomic paradigm on profit-maximizing private firms and then identify the deviations from that paradigm that are characteristic of state-owned firms. The works quoted below invariably follow that approach. The method used to establish the deviations consists of generalizations from more or less systematic empirical observation. In some studies, these generalizations relate to basic issues, such as corporate goals and associated organizational and management structure, and these are the starting point for logical deductions about state-owned corporate behavior. Other studies ignore the basic characteristics and concern themselves directly with empirical generalizations about behavior in such areas as investment, cost, and profitability.

Ideally, the basic characteristics and behavioral patterns of private profit-maximizing firms and state enterprises would be clearly distinguishable, but the studies reveal that the real-world conditions are quite blurred. Private firms seldom conform completely to the pure microeconomic paradigm, while the characteristics of state enterprises appear to range from forms quite akin to private corporations to those with the full characteristics of government agencies. Although the literature cites numerous exceptions to most of the statements on typical features of state-owned firms, the emergent message is that a difference between the average private firm and the average state enterprise does exist and can be identified.

Characteristics of State Enterprises

One commonly observed characteristic of a state enterprise is that its goals are less clearly specified than is usual with private enterprise. The multiple objectives of state-owned firms reflect conflicting public needs and political pressures.[1] These objectives may involve consideration of externalities, enterprise or national employment, income distribution, regional equality, and national sovereignty. Existing conflicts and tradeoffs among the goals usually have not been fully sorted out.

Another feature typical of state enterprise is the hazy relationship between top management and the owners.[2] In many cases, the owners cannot be clearly identified and certainly do not speak with one voice. The state commonly exerts its ownership rights through a variety of individuals and institutions. Economic reasoning suggests that in such circumstances the owner representatives who happen to have the greatest influence at a particular point in time

1. Y. Aharoni, "Managerial Discretion," in R. Vernon and Y. Aharoni, *State-Owned Enterprise in the Western Economies* (London, Croom Helm, 1981); M.M. Shirley, "Managing State-Owned Enterprises," World Bank Staff Working Paper No 577 (Washington, World Bank, 1983).

2. Y. Aharoni, "Managerial Discretion." Also see D. Coombes, *State Enterprise, Business or Politics?* (London, Allen and Unwin, 1971).

will have substantial scope for extracting economic or political benefit to themselves by pressuring the state enterprise to interpret its objectives in a particular way and to transact its business accordingly.

The tendency is for state enterprise to operate under less stringent financial constraints than comparable private units.[3] State enterprises commonly benefit from outright reductions in the "normal" tax burden and from little pressure for dividends on equity.[4] Such arrangements obviously increase cash flow. Public enterprises also usually have access to capital on advantageous terms, which can take a variety of forms such as low-cost credit made available by the ministry of finance, government guarantees for loans obtained from private sources, or permission to issue tax-free bonds. On the other hand, the access of state enterprises to the purely commercial markets for risk capital is often restricted because of the statutes under which they are operated or because of their less stringent requirements about returns on capital. Thus, the supply of capital to state enterprises is more governed by political factors than by commercial considerations as compared to private profit-maximizing corporations.

In government, state enterprise regularly has a savior of last resort.[5] Even if governments have frequently saved large private enterprises from bankruptcy and provided support for their continued life, seldom has any presumption—or assurance—existed that such support would be forthcoming. In distinction, a state enterprise can be reorganized, merged, or sold, but in practice, it is rarely allowed to default. Undercapitalization resulting from unprofitable operation is regularly remedied through new financial infusions.

Related to the assurance of survival for a state enterprise is the survival assurance granted to its management. Top management tenure in publicly owned firms is not absolute, but typically it is much stronger than in private corporations. Contributing to this security are the formal conditions providing job security to public servants, the government protection against discontinuation of activities through default, and the great difficulty of establishing managerial failure when several goals are pursued at the same time. When managers are sacked, it is usually because they have fallen out of political favor or broken some formal rules for public employee behavior.

3. W.S. Vickery, "Actual and Potential Pricing Practices Under Public and Private Operations," in W. Baumol, ed., *Public and Private Enterprise in a Mixed Economy* (New York, Macmillan, 1980).

4. M. Gillis, "The Role of State Enterprises in Economic Development." Discussion Paper 83 (Cambridge, Mass., Harvard Institute for International Development, February 1980).

5. Ibid.

Behavioral Patterns

The five features characteristic of state enterprise identified in the literature—a complex and blurred goal structure, an unclear relationship between top management and ownership, favorable access to financial supply, a virtual survival guarantee by the government, and managerial security of tenure—have been used along with direct empirical observation to derive several generalizations about state enterprise behavior.

First, the combination of blurred objectives and unclear ownership roles gives management much greater discretion in determining enterprise objectives than is common in the private sector. The complexity of goals and the variety of views among owner representatives on the weights to be attached to each often allows management considerable freedom in choosing the goals the state-owned enterprise will pursue. Managerial discretion is further enhanced by the virtual impossibility of measuring performance when goals are complex and the desired tradeoffs among them remain unidentified.[6]

Second, and consequent to the lack of managerial accountability, state enterprises commonly are under less pressure to minimize costs than are private firms because profit maximization is not the overriding objective of enterprise activities.[7] The pressure to minimize costs is particularly low in the many natural monopolies, in which state enterprises are common. More generally, high costs, or even an inability to cover costs with the combination of sales revenues and subsidies can always be justified by the pursuit of some social objective. An inadequate cost performance can always be excused by citing both legitimate and illegitimate reasons for the high costs, especially when the two cannot be clearly distinguished. The knowledge that in the end almost any losses will be covered by the government to assure survival adds to the permissive view on costs in state enterprises.

A third behavioral pattern deduced from the features that characterize state-owned firms and confirmed by direct observation is that they operate under bureaucratic systems. In such bureaucracies, adherence to formal rules becomes more important than goal achievement. Maximization of budget support or of operations replaces the maximization of profit, and decision making

6. Y. Aharoni, "The State Owned Enterprise: An Agent Without a Principal," in L. Jones et al., *Public Enterprises in Developing Countries* (Cambridge, Eng., Cambridge University Press, 1982).

7. H. Leibenstein, "X-Efficiency and the Analysis of State Enterprise." Paper presented at the Second BAPEG Conference on Public Enterprises in Mixed Economy LDC's, Boston, April 1980.

and decision implementation tend to be slow.[8]

Bureaucratic characteristics would be most evident in those state enterprises that operate in shielded markets without competition. The emergence of such characteristics can be seen partly as a result of the control and supervision problems the owners experience. When performance is difficult or impossible to measure, the tendency is to impose bureaucratic rules and procedures to assure that the activity does not go completely astray. The bureaucratic climate tends to lead to self-selection in the managerial cadres. Efficient managers want to be judged and rewarded on the basis of their performance. When performance is difficult to measure, managerial rewards will be based on other criteria, resulting in an exodus of the most efficient among them.[9] Those who stay will, on average, have less cost consciousness than managers in private enterprise.

The security of having government backing and management tenure in state enterprise has led some authors to conclude that those executives would be more willing to take risks than executives in private firms. Even though instances of extreme risk taking can be cited (for example, Pertamina in Indonesia), the bureaucratic tendencies and the ensuing managerial selection process pull in the opposite direction so that the managements of state enterprise are often risk averse. No clear-cut conclusion about the net impact of these opposing tendencies can be drawn from available empirical evidence.[10]

Based on these behavioral patterns, a reasonable expectation would be that state enterprises are less profitable than private enterprises in similar circumstances. Empirical observation supports this expectation.[11] The depressed profitability depends on two factors, however. The first, and desirable one, is that state enterprises may be performing some useful social function that is not fully credited in their accounts. The second, and undesirable one, is that these enterprises usually operate at a lower level of internal efficiency.

8. L. Jones and E.S. Mason, "The Role of Economic Factors in Determining the Size and Structure of the Public Enterprise Sector in Mixed Economy LDCs," in L. Jones et al., *Public Enterprises in Developing Countries* (Cambridge, Eng., Cambridge University Press, 1982); W.A. Niskanen, *Bureaucracy and Representative Government* (Chicago, Aldine Atherton, 1971); and A. Downs, *Inside Bureaucracy* (Boston, Little Brown, 1967).

9. Y. Aharoni, "The State Owned Enterprise."

10. M. Gillis, G.P. Jenkins, and D.R. Lessard, "Public Enterprise Finance in Developing Countries: Towards a Synthesis," in L. Jones et al., *Public Enterprises in Developing Countries* (Cambridge, Eng., Cambridge University Press, 1982).

11. H.G. Can and G. Dutto, "Financial Performance of Government Owned Corporations in LDC's" (Washington, *IMF Staff Papers*, March 1968). Also see evidence contained in M.M. Shirley, "Managing State-Owned Enterprises."

A fourth behavioral pattern among state enterprises is that production is more capital intensive than in private enterprise.[12] This follows partly from the observation that state enterprise tends to be concentrated in capital-intensive sectors of the economy.[13] Even when the same types of activities are compared, however, the level of capital intensity would be higher in the state-owned units. The heavy use of capital can be seen as a rational adjustment by state enterprises to the low cost they pay for it.[14] Some authors argue that the high capital intensity is in part the result of management's attempt to increase its discretionary power; the owners' ability to influence and direct is greatest when the firm requires additional capital infusions to cover operating losses.[15] A high capital intensity, reducing variable cost levels, will diminish the likelihood of operating losses. The need to apply to the government for additional funds will consequently be less frequent. Often, politicians are quite willing to play this game. Decisions about additional capital allocations to rescue or expand public enterprise activities are easier to make and carry more political prestige than the painful and laborious task of assuring more efficient use of the resources already at the state-owned firms' disposal.

A fifth behavioral feature of state enterprise, closely related to the one just treated, is that capacity will be greater than private enterprise would establish under similar circumstances. This follows from the low cost of capital and also from the bureaucratic tendency of management to maximize the volume of operations rather than profits.

In summary, then, the economics literature surveyed brings out the following behavioral patterns typical of state enterprises:

1. substantial managerial discretion in delineating enterprise goals,
2. low pressure to minimize enterprise costs,
3. bureaucratic organization,
4. high capital intensity, and
5. overinvestment in capacity.

12. M. Gillis, "The Role of State Enterprises."

13. W. Baer and A. Figueroa, "The Impact of Increased State Participation in the Economy on the Distribution of Income." Paper presented at the Second BAPEG Conference on Public Enterprises in Mixed Economy LDC's, Boston, April 1980.

14. M.M. Shirley, "Managing State-Owned Enterprises."

15. M. Gillis, G.P. Jenkins, and D.R. Lessard, "Public Enterprise Finance."

Behavioral Patterns of Mineral Enterprises in Developing Countries

The "pure" models of state-owned and of private mineral enterprises identify several major behavioral features that distinguish state-owned mineral enterprises in developing countries from the private mining multinationals. In reality, many mineral enterprises combine traits from both models.

The Private Multinational Firm

The overriding objective of the private multinational mineral firm is long-run profit maximization subject to avoidance of excessive risk. This objective is tolerably clear-cut and can be interpreted in operational terms without undue difficulty. The goal of profit maximization obviously does not preclude pursuing social objectives required by general law, workers' safety, for example, or those, such as employee training, related to longer term profitability.

The lines of command and the relationship between management and owners in multinational mining corporations are commonly reasonably straightforward. Management is hired to ensure that the firm attains its objectives. Remuneration of managerial personnel and its security of tenure are related to profit performance. This assures a downward pressure on costs. The firm will have acquired considerable experience in mining and mineral processing long before it developed into a multinational. Its personnel have the expertise required for efficiently developing mines, running mineral production operations, and worldwide marketing of the output. For these reasons, inefficiencies internal to the firm would usually be limited, and most of the feasible mineral rent would emerge as before-tax profit. Because the firm is multinational, however, its profits and the activity from which they are derived may be geographically separated.

The objective of profit maximization implies capacity utilization up to the level where marginal cost equals price. Capacity utilization will be high during booms, when prices are high, and lower during recessions, because of falling prices. The higher the proportion of variable cost to total cost, the greater will be the optimal output reduction in response to a fall in price.[16] Quite frequently, the structure of the mineral industry is oligopolistic, and major producers have some tacit understanding among themselves that all will benefit from production cuts during recessions. In such cases, a multinational

16. F.M. Scherer, *Industrial Market Structure and Economic Performance* (Chicago, Rand McNally, 1980) pp. 205–209.

mining firm would adjust output even further downward than it normally would, to the level at which marginal cost equals a marginal revenue schedule derived on the assumption that its major rivals adjust output similarly.

Curtailed capacity utilization involves a proportional reduction in all variable costs, except if the recession is deemed to be temporary and the startup costs of certain functions are so high that it is more economical to maintain those functions throughout the recession. For instance, unskilled workers may be more readily dismissed than skilled personnel, who may be maintained because of the investment in their training and the risk that they may not return once they have been fired.

The flexibility of the multinational firm to adjust capacity utilization in response to variations in demand may be reduced by government legislation that restricts the freedom to lay off workers or by selective government subsidies to keep the capacity fully operational. Such government measures will be most common where the mineral activity is economically important to the region.

A multinational firm's investments are constrained by its financial capacity. A heavy investment program will strain its financial resources, increase its debt-equity ratio, and raise the cost of obtaining additional debt financing. In addition to resisting this rise in the cost of financing, the firm will be reluctant to allow its debt-equity ratio to increase above some threshold level because of the mounting risk of financial failure.

Potential mineral investment ventures with expected after-tax discounted cash flow (DCF) rates of return on equity above some "normal" return on capital in industrialized countries will be ranked according to their respective return levels. Priority will be given to those with the highest after-tax return, with some side conditions to satisfy the avoidance-of-risk criterion. For instance, in calculating the rate of return for individual ventures, the company may adjust downwards the anticipated revenue from projects in developing countries to account for the political risk of nationalization, of a sudden and drastic change in the fiscal regime, or of other political measures that could reduce its production and supply capability. The higher the perceived risk, the greater would be the downward adjustment. At some perceived political risk level, potential ventures would not be considered at all, regardless of their anticipated return.

To reduce the detrimental implications to itself of the political risk factor, the firm will also favor a wide geographical spread of its investments in developing countries. As a result, multinationals will undertake little or no investment in developing countries with a favorable mineral potential where taxes are high, where political conditions are deemed unstable, or where the company already has substantial investment involvements. The company would tend to make an upward adjustment in the anticipated revenues from projects that increase its vertical integration, however, because this involves the benefit of securing downstream markets and reducing the commercial risk of marketing. This may explain the prevalence of vertical integration in multinational mining firms.

In oligopolistic international mineral markets, a multinational firm will, as a rule, also constrain investment because of a tacit understanding that its competitors are similarly constrained in a joint effort to restrict long-run supply to a level at which prices cover costs and provide an adequate return on equity to the marginal venture. This, of course, does not preclude aggressive behavior by individual firms in the process of expanding their market shares.

Even in the absence of such tacit understanding, the expansion of the industry as a whole will be constrained by financial institutions concerned about problems of loan repayment in a situation of oversupply and low price.

The Mature State Mineral Firm

A mature state mineral firm is defined as one that has been in business for a long period of time and that does not suffer from inefficiency due to inexperience. Examples of mature state-owned mineral corporations include Companhia Vale do Rio Doce (CVRD) in Brazil, Codelco in Chile, and PT Timah in Indonesia.

OPERATIONS. Like the private multinational firm, the experienced state-owned corporation would have built up an in-house availability of technological and managerial competence in operating existing mineral projects and in developing new ones. The availability of such competence, however, does not assure a degree of efficiency equal to that found in private enterprises. The earlier finding that features typical of state enterprise—uncertain goals, security of management tenure, and unclear relationship between owners and management—make this firm less efficient and less profitable than private enterprise applies fully to state-owned mineral enterprises in developing countries.

Assuming that economic and social progress is the overriding goal, the national government could control the mineral firms under its ownership in either of two ways. One approach would be to require the firms simply to behave like private companies and to maximize their profits, while attending to social objectives only when required by law or to improve long-run profitability. The government could then spend the revenues from the mineral sector to accomplish its social benefit goal in ways it thought most fit. An alternative approach would be to require the mineral firms to deviate from the profit maximization objective and to take on wider tasks external to the company's operation in an effort to contribute more to the nation's economic and social progress. The rationality of the latter approach is contingent upon the suitability and efficiency of the mineral firms as instruments to pursue broader objectives.

In practice, governments in developing countries almost invariably choose the latter approach. Judging from official statements, state-owned mineral firms are required to pursue the profit objective and simultaneously to give special consideration to the direct and indirect contribution of their operations to the nation's foreign exchange income, to national value-added, to employment and skill creation, to technological proficiency, and to general industrialization in the national economy.

To the casual observer, the objective of the state mining firms might seem an inconsistent pursuit of maximizing these goals simultaneously. On further reflection, however, such multiple goals, at least ideally, can be fully consistent with the single objective of contributing as much as is feasible to national development or, in the economists' jargon, of maximizing the social rate of return on the nation's capital outlays for mineral activity.

To demonstrate this point, regard the maximization of profits as a first rough approximation of the national benefit objective and consider the adjustments needed to make it a truer measure of national benefit. Prices are often distorted in developing economies. The rationing of foreign exchange and unemployment suggest that the socially optimal, or shadow price, of foreign exchange is higher and the shadow wage for labor lower than actual market rates. Under such circumstances, the pursuit of national benefit will require that greater importance should be attached to earning foreign currency and to employment than when the goal pursued by the firm is simple profit maximization. The concern about creation of skills and technological proficiency, even beyond that motivated by profit maximization, is warranted by the positive externalities of these factors. The insistence on local inputs or domestic processing, even when privately unprofitable, may be a rational approach in the dynamic pursuit of national benefit in order to help establish a broad and diversified industrial base in the nation.

In principle, all these considerations could be caught in a dynamic programming model that would specify the operational modes needed for maximizing the social rate of return from the company's operations. In practice, governments adopt the more pedestrian approach of pursuing several goals simultaneously without specifying in detail the tradeoffs among them.

The rational pursuit of national benefit maximization would imply that a state company forgoes some private profit opportunities in order to attain a higher social rate of return. Hence, it would be wrong to judge the company's performance solely on the basis of the private profit criterion.

A number of problems and complications will cause reality to differ from this ideal and will reduce the efficiency of the state-owned mineral firm in gaining both social returns and profits. First, the government can choose among a number of policy measures for inducing or forcing the state-owned mineral enterprise to behave so as to maximize the social rate of return, and some of these measures will be less efficient than others in attaining the desired objective. General controls, such as a prohibition against firing workers or a quantitative allocation of foreign exchange, would usually involve a loss in efficiency, because the inflexibility of these controls would prevent the company from adjusting to the socially optimal position. Taxes and subsidies affecting the costs and prices of the company would be superior in this respect.

A second problem arises when government policies are not well adapted to the purpose of maximizing national benefit. For example, providing inputs, such as capital or energy, to publicly owned units at rates lower than their social opportunity cost may provide faulty signals to the state-owned mineral firm.[17] The question is then whether the state-owned companies should disregard the signals resulting from misdirected public policies and work instead toward what they perceive as "true" social goals.

A third problem is that the national benefits resulting from the state-owned companies' activities are exceedingly difficult to quantify. The ensuing unclarity of company goals makes judging managerial performance difficult. This unclarity, along with the role the government assumes as savior of last resort against threats of bankruptcy, reduces the pressure on management to minimize costs.

Similar effects follow from the diffused ownership structure commonly found in state-owned mineral enterprises. Different ownership interests, such as individual ministers or provincial governors, often can impose their wills on the enterprise and distract its attention from national social benefit.

17. For instructive examples, see M. Gillis, "Allocative and X-Efficiency in State-owned Mining Enterprises" (Cambridge, Mass., Harvard Institute for International Development, April 1979).

Empirical observations illustrate that these inefficiencies have taken a variety of forms. Both management and owners can get away with policies aimed at increasing their own private benefit by claiming that such policies are needed to maximize the state firm's contribution to social welfare. Leisure facilities for all employees, but in practice used predominantly by the top management, belong to this category. The excessive use of the company to promote regional development can be presented as having a valuable social objective, even though the major purpose may be to create a political power base for local politicians. An unnecessarily large labor force or extra capital installations can make management's job easier, but the multiplicity of goals would make it difficult to determine what the socially optimal levels should be. Inefficiencies of this kind are likely to be especially common in projects with particularly large mineral rent potential, in which profitability will be deemed acceptable even after management and owner interests have taken their share.

In summary, therefore, the combination of empirical observations and economic reasoning presented here suggests that the way state mineral enterprises use the resources at their disposal is subject to substantial inherent inefficiencies whether their performance is measured in private profits or a broad set of social objectives.

Inefficiencies apart, an important consequence of the broader goals for the operating modes of state-owned mineral firms will be some inflexibility in capacity utilization over the business cycle. A corollary of this inference is that prices will fluctuate more in markets with important state enterprise presence, because of the lesser adjustment of output to changing demand.

Some clearly variable costs will prove to be fixed when the social rate of return rather than private profit is the firm's objective. This may, for instance, be true of the cost of labor if other jobs are not available. With a lower share of costs in the variable category, the state-owned enterprise will be less likely to reduce its capacity utilization when prices fall, because its variable costs are so low that they will be covered even at very low prices.

If the shadow rate of foreign exchange lies above the private rate, the social revenue from export sales will be higher than private revenue; for this reason, the state-owned firm may find it rational to continue full-capacity operations when a private multinational would not. In the common case, the state-owned mineral firm is a dominant foreign exchange earner, so a decline in demand and price will tend to increase the differential between the shadow and official rates of exchange. The reduced mineral earnings will make foreign exchange scarcer and hence increase its shadow price in local currency. Expressed at shadow rates, the value of a dollar earned will then be especially high when mineral export revenues decline. The result will be to further reduce the dollar price of minerals at which capacity reductions become socially warranted.

Such implicit or explicit consideration of externalities, whether imposed by government or management, will tend to make state-owned mineral firms less willing to reduce capacity utilization in consequence of declining demand and price and more eager to resume full capacity when prices recover.

Formal testing of the contention that state mineral enterprises are less flexible than private ones about capacity utilization is difficult. Defining output instability as the percentage deviation in output from its five-year moving average, L. S. Powers concludes that developing countries on average have been somewhat more stable than industrialized countries in output of copper, aluminum, tin, and iron/steel in the 1950–77 period.[18] She hypothesizes that this difference might have to do with the greater state involvement in these industries in developing countries. The Powers analysis is far too general to address the issue presently in focus, however. First, it does not specify the countries and periods the respective mineral industries were under public ownership. Second, the output instability measure provides an aggregate reflection of all the forces that induce output changes and does not isolate the impact of price variations on output. Third, as admitted by Powers, the method used yields a high measure of instability for countries that have experienced rapid, large, stepwise unit output increases as new projects have gone into production. This type of output change does not relate to the ability to adjust utilization of existing capacity to changing market circumstances.

Ideally, a test to verify the thesis that state mineral enterprises are less flexible than private firms in adjusting capacity utilization to falling demand would require detailed sets of data on capacity for individual firms or at least individual countries at each point in time. This would permit a determination of the extent of variation in capacity utilization over time. Though capacity is a vague and fluid concept, occasional studies of mineral industries provide detailed capacity figures for an individual year. Tables 2-A-1 through 2-A-6 are based on such studies. Comprehensive and reliable data of this kind covering a long period of time have not surfaced, however.

The present study's not very successful attempt to verify the hypothesis about output inflexibility, like that of Powers, relies on comparing variations in production for private and public entities. Table 3-1 shows output adjustments for six copper-producing countries in which the industry is predomi-

18. L.S. Powers, "Instability in the Copper, Aluminum, Tin and Iron and Steel Markets," *Materials and Society*, no. 3, 1981.

Table 3-1. Downward Adjustment in Copper Mine Production, 1974–82
(percentage change from preceding year)

Country and group	1974	1975	1976	1977	1978	1979	1980	1981	1982
Countries with predominantly public ownership									
Chile		-8.2			-2.8				
Finland	-3.9					-12.2	-10.5		
Mexico		-5.1			-2.8				
Yugoslavia				-3.2		-9.7		-5.0	
Zaire		-1.0	-10.1		-12.0	-5.4			-2.0
Zambia	-1.2	-3.0		-7.5	-2.8	-8.6		-1.4	-9.8
Unweighted average	-0.9	-2.9	-1.7	-1.8	-3.4	-6.0	-1.8	-1.1	-2.0
Countries with predominantly private ownership									
Australia		-12.8						-8.2	
Canada		-10.7	-0.4		-13.2	-3.5			-10.0
Philippines					-3.7			-0.7	-5.9
South Africa					-3.1			-0.6	-1.6
Sweden	-9.4					-3.3	-6.6		
U.S.	-7.1	-11.5		-6.3			-18.2		-26.2
Unweighted average	-2.8	-5.8	-0.1	-1.1	-3.3	-1.1	-4.1	-1.6	-7.3
Copper prices (constant 1981 U.S. cents/lb)	166.2	87.1	97.0	83.5	73.5	95.6	95.6	79.0	67.5

Source: Metal Statistics, 1983.

nantly publicly owned and for six others in which ownership is almost exclusively in private hands. The countries included accounted for more than 80 percent of Western World copper mine production in recent years. Only the period from 1974 to 1982 is covered. In the 1964–74 decade, copper prices were so high that few mines failed to cover their variable costs. In addition, many of the copper mine nationalizations did not occur until the seventies.

The table shows percentage decreases of output compared with the preceding year for each country. Unweighted averages of the decreases for each country group are also given. Increases in output are ignored on the presumption that they are dominated by expansion of capacity, which is not a concern in the present investigation.

The average annual downward adjustment for the whole period is indeed greater in the private group (−3.0 percent) than in the public group (−2.4 percent), but the public group shows larger downward adjustments in four out of the nine years. Zaire and Zambia are the two countries in the public group that contributed most to the downward adjustment figures during the period studied. Demand and price conditions played a minor role for the output reductions in these two countries, however. Political upheavals in the Shaba province mainly explain the recorded output reductions in Zaire, and the negative figures for Zambia in seven of the nine years covered reflect a falling trend in production capacity.

A study of negative deviations from trend instead of downward adjustments from the preceding year would better explain the Zambian situation, but this method would also introduce two other complications. First, it runs into the difficulty experienced by Powers—large, step-wise output growth produces negative trend deviations. For example, copper output in Mexico jumped upward in 1980, when the large La Caridad mine went into production. As a result, in the 1975–79 period, Mexico would appear to have sizable negative deviations from trend output. These deviations had little to do with demand and price in the international market.

Another problem with using deviations from trend to compare output flexibility is that in five of the twelve countries studied the computed trends (obtained on the basis of least-square deviation) during the 1974–82 period are not significant. The significance of trends might be improved by deriving them from data for a longer period of time, but this too poses problems because of the sharp and sudden deceleration in world economic growth about 1974 and the ensuing kink in the trend expansion of copper demand and supply.

Some of these problems could be overcome by concentrating on the years of significant price falls. The rationale is that downward output adjustments in other years cannot have been caused by falling demand and price. The price series given in table 3-1 shows 1975, 1978, 1981, and 1982 as the years that warrant special attention. Even then, the complications of political upheaval in Zaire and the downward trend in Zambia's copper production remain. Furthermore, the three-year schedule for U.S. copper industry labor negotiations caused strikes and sizable output reductions in 1974, 1977, and 1980. The downward adjustment of U.S. copper output to the fall in prices in 1975 would have been much stronger in the absence of the strike-induced contraction of 1974. Similarly, output curtailment appears to have preceded the price reductions of 1978 and 1981 and to be related to the strikes of 1977 and 1980.

Even when considering only the years when prices fell, a key factor in the intercountry variation in downward adjustments of output will be the level of variable costs in each country's copper industry. Thus, the low production costs in Australia as compared to Canada probably provide the major explanation for the lesser output adjustments in Australia.[19] Cost levels probably also

19. A.F. Barsotti and R.D. Rosencranz, "Estimated Costs for the Recovery of Copper from Demonstrated Resources in Market Economy Countries," *Natural Resources Forum*, No. 2, 1983.

explain the variation in Chile's experience. Production was cut in 1975 and 1978 but not in 1981 and 1982 because dollar costs were substantially lower in the 1980s as a result of exchange rate adjustments. Although instances of this kind can be identified, isolating the impact of cost levels on the flexibility of capacity adjustments in each country is not feasible because of deficiencies in the cost data.

Formally establishing the relationship between the emergence of state enterprises and price instability is even more difficult than verifying and quantifying the impact of state enterprise proliferation on output flexibility over the business cycle. The logic in support of such a relationship is clear, but the empirical data surveyed and the analytical tools available at the present time do not permit definitive proof that the proliferation of state mineral enterprises on a large scale has destabilized international mineral markets.

A working hypothesis formulated in the introductory chapter is that experienced state-owned firms adopt an oligopolistic conduct quite similar to that of the private mining multinationals. So far, the analysis of state-owned enterprises' characteristics and behavioral features does not warrant a rejection of that hypothesis. The experienced state-owned firm will be as aware of the interdependence among sellers as are its private counterparts, and its reactions to this interdependence are likely to be similar. Although state-owned and private firms may differ in evaluating their marginal revenue and marginal cost schedules, a large state-owned firm in a concentrated industry, like a private one, will appreciate the downward sloping demand for its output resulting from the interdependence and from the ensuing tacit collusion among suppliers in the world market. The emergence of producer government associations, such as the International Bauxite Association (IBA) in bauxite, CIPEC in copper, and Association des Pays Exportateurs de Minerais de fer (APEF) in iron ore, testifies to the awareness of the exporting governments and of their enterprises about international interdependence among suppliers. Given this awareness, and abstracting from ideological inhibitions, the goals of state-owned enterprises will require that marginal cost be equated with marginal revenue when determining how much to produce, even though the marginal schedules of the state-owned firms are derived on the basis of social rather than private criteria.

INVESTMENTS. The broader goals and the inefficiencies that attach to the operations of state-owned mineral enterprises apply equally to the choice and execution of investment projects. Two further issues require additional investigation, however. First, how does state ownership affect project selection and hence the long-run structure of the industry worldwide? And second, does state ownership result in more or less capacity expansion than private ownership?

Private multinational firms and state mineral enterprises differ in how they evaluate and rank potential mineral investment projects in developing countries. Private firms will be guided by the after-tax, discounted cash flow (DCF) rate of return on equity, subject to a reduction in anticipated revenues to account for political risk and to a cutoff rate reflecting some "normal" return on capital in the industrialized countries where they operate.

If the assumption that economic and social progress are the overriding goals of the national government holds true, then mineral industries owned and run by the state should rank potential mineral projects according to the anticipated social DCF rate of return, with the entire (social) before-tax revenue considered as a benefit, that is, with no explicit downward adjustment for tax or political risk. The cutoff rate would reflect national capital scarcity; it would equal the rate of return that the same capital could earn in marginal alternative ventures. Even though exceptions should not be difficult to find—for example, in mineral ventures launched by the state primarily for national prestige reasons and disregarding economic returns—empirical observations suggest that the selection criteria outlined are common among state-owned mineral enterprises.[20]

How are investment patterns likely to differ between private and state-owned mineral enterprises? Consideration of the social instead of private rate of return should advance the ranking of more labor-intensive projects, those that use a larger proportion of domestic inputs, and those with smaller negative environmental consequences than the average of projects under scrutiny. Should few potential projects present themselves, however, and given the ambiguities of social project evaluation, the practical importance of this difference in ranking is uncertain.

Consideration of the entire revenue earned by state-owned enterprises, without adjustment for taxes or for political risk, is likely to increase substantially the rate of return, thereby augmenting the number of projects with expected rates of return that exceed any given cutoff rate. Rates of return would be especially attractive for state ownership in the case of projects with high potential profitability in which, as is common, fiscal levies are imposed on profits and in the case of countries where private multinationals experience a high degree of political risk. This is because the high profits tax and the

20. R. Mikesell, *The World Copper Industry* (Baltimore, Johns Hopkins Press, 1979).

substantial political risk to which private investors are exposed in such cases would induce them to make especially large downward adjustments in the rate of return for purposes of project ranking.

The cutoff rate of potential return state mineral corporations in developing countries apply should in principle be higher than that for multinationals because of capital scarcity in those countries. In practice, state enterprises may be justified in applying a rate similar to that used by the multinationals (for instance, the going international interest rate), because the marginal financial resources they need to develop their mineral projects are ordinarily obtained in the international market at such rates. Like any private multinational firm, a state-owned mineral company's investment program will be constrained by its ability to obtain financing. It will not develop all potential projects with rates of return above the cut-off. The resources of national financial institutions in developing countries are usually inadequate to meet the needs of modern, large-scale mineral projects.[21] Because of other pressing requirements for funds and because outside financing is more readily available for mineral projects than for, say, agricultural development, the governments of developing countries are commonly unwilling to make sufficient allocations in their budgets to meet a state mineral company's investment needs.[22] For these reasons, but also to assure their large foreign exchange requirements, state-owned firms—like private ones—rely heavily on financial sources in the international market, including multilateral agencies like the World Bank, for a large proportion of their investment funds.

In distinction from direct foreign investors, foreign lenders to developing countries have seldom been exposed to expropriation of their assets on political grounds. Hence, they have not shared the political risk perceptions of the multinational mining companies. Given the same commercial risks, lenders usually have been indifferent between the private multinational and the experienced state enterprise. In fact, for a long time, state enterprises may have had some small advantage in procuring financing when their owner, the national government, was considered credit worthy and was prepared to guarantee the loans. In broad terms, the limitations and constraints on the availability of investment funds have probably affected private multinationals and state mining companies quite similarly. The observation that state enterprises, in general, have superior access to finance compared to private multinationals ap-

21 See M. Radetzki and S. Zorn, *Financing Mining Projects in Developing Countries* (London, Mining Journal Books, Ltd., 1979).

22. Ibid.

pears to have had limited validity in the mineral sector in developing countries.

The international financial crisis that struck many Third World countries in the early 1980s sharply curtailed all commercial lending to developing countries. The risk of national default has turned the former advantage of the state-owned firms' link with their governments into a detriment. The precarious financial position of these nations has also discouraged lending to multinationals for mineral projects in developing countries. All servicing of old foreign loans has been put in jeopardy by the draconian restrictions on foreign payments in many of the financially distressed countries. New commercial lending for mineral projects in developing countries has therefore come to require special host government guarantees that loan servicing would have first option on the export proceeds from the project to be financed or repayment assurances, or both, from export credit agencies in industrialized countries if equipment delivery is involved. This policy has applied both to the investments of state-owned enterprises and to those of the mining multinationals.

Mineral investment activities in the long run are crucially dependent on exploration programs that identify new ore bodies and prepare them for exploitation. The ability of state enterprises to attend to this task efficiently has sometimes been doubted. Mikesell, for instance, claims that state enterprises do not command sufficient financial resources for exploration and that their national character prevents an efficient spread of exploration risks.[23] Mining company executives have suggested that state enterprises' exploration programs are likely to have a parochial outlook, focusing on what can be found nationally, without sufficient regard for what can be sold profitably in the international market. A comparison of groups of state-owned and private mining companies of comparable size reveals that, in the 1970s, the state-owned enterprises spent much less on exploration than the private firms did.[24]

The relative efficiency of mature state-owned enterprises' exploration programs remains an open question. Exploration normally accounts for no more than 5–10 percent of total investments in mining enterprises, so the state-owned enterprises should be able to command sufficient financial resources to cover their exploration activity. To generalize from the observation that a group of state-owned enterprises spent relatively little on exploration in the 1970s is hazardous. The firms may have limited their exploration because they

23. R. Mikesell, *New Patterns of World Mineral Development* (London, British-North American Committee, 1979).

24. P.N. Giraud, *Geopolitique des Resources Minieres* (Paris, Economica, 1983).

considered their mineral reserve positions adequate at the time or, more probably, because several of the firms were new and inexperienced. A judgment about the adequacy of exploration must also take into account the support provided by international agencies, such as the United Nations Development Program, whose expenditures may not be recorded in the state firms' accounts.

The argument about inefficient risk spreads may have some validity, especially for state enterprises in small developing countries. Note, however, that small and highly specialized exploration firms with little scope for spreading risk have proved quite efficient and successful.

The similarity of oligopolistic conduct in the operations of mature state-owned enterprises and those of private multinationals should, by analogy, apply in equal measure to investments. Thus, the investment behavior of both in an oligopolistic world market would be constrained by a tacit understanding among suppliers that profitability will suffer if production capacity is expanded in an uncoordinated manner. State-owned firms, like private multinationals, may have periods of aggressive behavior during which they expand their capacity and market share without much consideration for the global consequences to the industry, provided they are able to finance the expansion. CVRD in Brazil through the 1970s and Codelco in Chile in the early 1980s provide examples of such behavior.

The establishment of state enterprises through nationalization has frequently involved ruptures of international vertical integration chains. Industries with a fast-growing state enterprise presence have therefore had a lesser degree of vertical integration than would have been the case under a continued private multinational regime.

The reduced degree of vertical integration is likely to remain a permanent feature. Although the ruptured international links will be replaced to some degree by more forward processing by the state mineral company in its home country, the extent of this national vertical integration is unlikely to match that under multinational arrangements. Locating certain stages of production close to the final consumers in industrialized countries has strong commercial advantages, and also many countries have inhibitions against large, direct foreign investments.[25] In some instances, state mineral enterprises in developing countries have become multinational by investing in processing facilities abroad; examples include Codelco in Chile, Gecamines in Zaire, and ZCCM

25. R. Vernon, *Two Hungry Giants, The US and Japan in the Quest for Oil and Ores* (Cambridge, Mass., Harvard University Press, 1983).

in Zambia, which have acquired important ownership positions in French and West German copper-fabricating plants. These instances notwithstanding, the social objectives of state mineral enterprises in developing countries tend to make them more nationally oriented and hence less prone to undertake multinational operations than are large private mining concerns.

Less vertical integration increases the number of arms-length rather than captive transactions. Economic reasoning suggests that broader arms-length markets should stabilize prices in those markets. Both buyers and sellers can accept extreme price levels without endangering the viability of their operations if the volume of transactions at these prices is a very small share of their total market exposure. As the arms-length market expands, however, price swings will be subdued because the price extremes will endanger both buyers and sellers if the transactions involve a sizable share of total trade, and they will become increasingly reluctant to accept the full range of price variations.

Reduced price instability in an expanding arms-length market does not necessarily imply a greater sales price stability for an individual mineral producer. The price swings in the arms-length market may abate, but nevertheless they may vary much more than the price movements under the captive market arrangements. The instability of the prices a producer receives may well increase as it transacts a growing share of its output in the arms-length market.

The supposition that price stability improves as marginal arms-length markets are widened is difficult to prove. In fact, instability in such markets could have increased despite the stabilizing impact of market widening. Growing synchronization of the international business cycle since the 1970s could account for greater price swings. Also, the unwillingness of the state-owned sector to cut capacity utilization in response to demand declines could have outweighed the price-stabilizing effects of market widening. A further problem is that the price movements in marginal arms-length markets prior to vertical integration ruptures are incompletely documented. For example, little systematic evidence exists of market prices for internationally traded iron ore before 1960, apart from the series for Swedish sales to Western Europe, which is somewhat special because of the close and long-standing relationship between the supplier and the buyers. Even in the early 1980s, only scattered evidence of arms-length pricing of bauxite is available. The two bauxite price series compiled by the World Bank in 1982 are (a) the U.S. import price, which presumably must reflect mainly the accounting prices of the U.S. aluminum multinationals; and (b) a series derived from Jamaican production and transport costs, which strictly speaking is not a price series at all.[26]

What conclusions can be drawn about how the increasing presence of state enterprise in developing countries' mineral industries has changed mineral investment patterns? The supposition that state enterprises disregard fiscal burdens and political risks in their assessments of returns from potential projects suggests that state ownership might lead to a shift of investment flows toward the mineral-rich developing countries that multinationals avoided because of severe fiscal regimes or perceived political instability. If so, the bias against mineral activities in certain developing countries with favorable economic potential for mineral exploitation that would have followed from a continued dominance of the private multinationals in global mineral activities should be reduced. This conclusion will only hold, however, if the increase in perceived returns resulting from the state enterprises' method of evaluating projects exceeds any possible reduction in returns due to less efficient management of exploration programs and higher costs of operating mineral ventures that are typical of publicly owned firms.

This analysis of mature state enterprise investment behavior does not suggest that the increasing importance of such enterprises will affect the global level of mineral investments. With existing interdependencies among producers in the mineral industries, any tendency toward expanded investments in certain developing countries would induce corresponding reductions elsewhere, unless investors in the marginal projects are prepared to accept a lower rate of return. Furthermore, the financiers would be reluctant to extend loans for expanded global investment programs that would reduce mineral prices and hence the ability of the new projects to service their loans.

The New and Inexperienced State Mineral Firm

The rationale for devoting ample space to the analysis of the inexperienced state-owned mineral firms follows from the extended period required for such firms to mature and from their peculiar behavioral features during this interim period. Given the recency of their establishment, relatively few state-owned firms have yet attained maturity. Through the 1960s and 1970s, a majority of all public mineral enterprises in developing countries were in the inexperienced category.

Inexperienced state-owned firms are of two types. One includes a majority of the companies set up to manage existing mineral ventures at the time of

26. Personal communication with the Commodities and Export Projections Division of the World Bank.

nationalization. The transfer of managerial responsibility from the private multinationals to such firms was usually sudden and has had profound implications for the industry. The second type includes mineral ventures launched after independence and in which the developing country governments had an involvement from the outset. The emergence of this type of enterprise has been much less conflict-ridden than was the case for the first type. The mining multinationals often remained involved through new investment arrangements such as joint ventures, production sharing, or lease contracts.

Many of the behavioral features of the new and inexperienced state mineral enterprises differ from those of the mature ones. A reasonable working hypothesis is that inexperience is a temporary phenomenon and that as the new firms gain experience their behavioral characteristics will gradually converge towards those of the mature state-owned mineral enterprises.

STATE FIRMS TO MANAGE EXISTING VENTURES. Those firms set up to manage existing nationalized operations usually had a difficult start. The former owners in many cases were dissatisfied with the compensation they were offered and were reluctant to provide managerial assistance and technology for the new national entities. The state firms therefore regularly took on wide-ranging responsibilities long before they could acquire the expertise necessary for the tasks at hand. The maturation period is characterized by substantial and extended set-up costs.

The normal activities required of mineral companies differ greatly in degree of complexity. The more complex the task, the longer will be the time required to establish the expertise to handle it efficiently. Factual evidence suggests that developing commercial skills takes less time than for technical operations. On the technical side, simple open-pit operations are easier to handle than the more complex underground ones. Management of investments to develop new projects, especially on a large scale, seems to be the most difficult task of all. Very few state mineral enterprises in developing countries have yet acquired the competence to carry out such tasks efficiently.

When a newly set up and inexperienced state-owned firm is suddenly put into a position of overall responsibility, the result is almost invariably disruption of operations. The resulting inefficiency must be distinguished from the organizational slack in mature state-owned firms. The latter arises from the fact that management or owners place other goals ahead of efficiency. The present problem follows from the fact that the appropriable yield of the firm—the residual after normal payment for factor inputs—is reduced by lack of managerial competence. Hence, less can be accomplished toward meeting the firm's goals, whether they be private profit, social benefit, managerial convenience, or the vested interests of some influential group.

A reasonable generalization based on scattered empirical evidence is that disruption reaches a maximum soon after takeover and then gradually subsides over time. Initially, all key aspects of the state mineral firm's operations are affected. Commercial relations with customers and competitors are frequently upset, especially if the state takeover ruptures a chain of vertical integration, thereby depriving the nationalized unit of its captive markets. The quality of products deteriorates. The inexperienced management is often unable to maintain production at full-capacity levels. Input requirements per unit of output tend to increase. Capacity expansion comes to a standstill.

As the new management gains experience, order is gradually restored and inefficiency losses decline. Commercial operations are usually the first to be normalized, though often along different patterns than before nationalization. If the nationalized units had been part of vertically integrated private multinationals, arms-length relations are developed in place of formerly internal transactions. Full capacity utilization is eventually restored. Excessive cost levels are reduced. Last of all, the company becomes able to expand its production capacity. The duration and gravity of the disruption will be related to the economic development of the country. It will be most severe in the least developed countries, where the foreign-owned mineral activity had been carried on within an enclave, with limited spillover to the rest of the economy. In many cases, ten to twenty years may be needed before the new firm attains full maturity.

Experiences of nationalized enterprises suggest that the initial disruption of commercial relations with competitors commonly includes a disregard for the oligopolistic collaboration that may have restrained global output and capacity expansion. This disregard is usually of no practical consequence, however. The initial inefficiencies of the state firm commonly constrain its capacity utilization and capacity expansion more than would any tacit or overt collaboration with members of a world oligopoly. By the time the new state firm becomes a mature and experienced corporation, it is likely to have overcome its inhibitions or inability to participate in any oligopolistic coordination.

Economic reasoning explains why inexperienced state-owned firms seldom venture to develop new mineral projects during their early life.[27] First, their

27. The Iranian government's decision to manage the development of the Sar Cheshmeh copper mine without any direct multinational involvement is an exception. The effort ran reasonably successfully until the fall of the Shah, at which time further work to complete the project was arrested for several years.

inexperience would cause substantial inefficiencies in investment, which could turn even very rich mineral deposits into privately and socially unprofitable operations. Second, the international credit institutions that finance mineral investments would be reluctant to extend loans out of fear that an inexperienced state mineral company might fail to complete its development plans. Third, the companies are likely to be so absorbed in mastering operational intricacies that for a long time they will be reluctant to undertake new and demanding tasks with which management has no prior experience. Not until after they had enough experience to run their operations smoothly and efficiently would the firms ordinarily feel in a position to invest in new projects.

One common way to reduce inefficiency losses after nationalization has been to hire a management team, either from the former owners or from another multinational mining firm. From the government's point of view, a management contract arrangement has several drawbacks. First, an element of monopoly exists in the supply of managerial services. Compensation of the management team therefore regularly absorbs part of the mineral rent. Second, even though in principle ultimate control remains with the government, if the nation lacks its own technically competent managers, it runs a serious risk that an experienced management team hired from a multinational company will attach less importance to social benefit goals than to its own interests. Some of the advantages of nationalization are lost as a result. Third, if foreign firms completely control management, the state mineral company will be reduced to a holding operation. Reducing the exposure of nationals to the mineral business proper has sometimes slowed down the process of building up national competence.

Although management teams have frequently been hired to prevent a collapse in existing operations, the governments have been very restrictive with regard to extent and duration of the hired teams' responsibilities. The teams are viewed as instructors and trainers of a national management cadre. The comprehensive but temporary use of management contracts has sometimes altered the sequence of postnationalization events by deferring the outbreak of disorder and inefficiency until after the departure of the outside managers.

If nationalizations involve a significant proportion of global capacity and are concentrated in time, prices would be expected to rise temporarily in consequence of the inefficient operations of the new state enterprises and their lack of investment. Also, the slowdown of capacity expansion in the nationalizing countries should be directly observable. Empirical verification is difficult, however, because state takeovers have usually been spread over time and because prices are usually affected simultaneously by many factors. Nevertheless, the exceptionally high copper prices in the 1964–74 period can be ascribed in some degree to the inability of Chile, Peru, Zaire, and Zambia to expand capacity after partial or total nationalizations of their industries and the ensuing conflicts between each of the four governments and the mining multinationals.[28] The average copper price in the 1964–74 period was $1.85/lb (constant 1981 dollars), 60 percent above the $1.15/lb average for the preceding eleven years. The shift to the higher price was related to the fact that the four CIPEC countries' annual average output expansion fell from 6.2 percent in the 1950–60 period to 2.7 percent between 1960 and 1974. Western World output outside CIPEC rose by an annual average of 3.6 percent between 1950 and 1960 and by 4.7 percent between 1960 and 1974.

STATE FIRMS TO LAUNCH NEW VENTURES. The second type of state mineral enterprise has emerged in the relatively recent past and under much more relaxed circumstances than did the first type. So far, the influence of these enterprises on behavior and modes of operation in the mineral industry has been very limited.

The establishment of the second type of firm did not involve a transfer of responsibilities for running existing operations, because none existed when the firms were created. Instead, the firms were set up to develop virgin mineral deposits into large-scale mining operations. Most commonly, they are joint ventures with one or several multinational mining companies. The government holds a significant or majority share of the equity, but the entire management function is entrusted to one of the private partners. Examples include the Companie des Bauxites de Guinee in Guinea, which became operational in 1973 and in which the government equity is 49 percent[29] and the uranium mining companies Somair and Cominak in Niger, operating since 1971 and 1978, respectively, in which the government holds 33 percent and 31 percent of the equity.[30] In petroleum, gas, and coal, the arrangements have often involved full government ownership, combined with a lease contract or production-sharing agreement with the foreign exploiter, but such arrangements are rare among nonfuel minerals enterprises.

28. For a discussion see M. Radetzki, "Long Term Copper Production Options of the Developing Countries," *Natural Resources Forum*, no. 1, 1977.

29. M. Radetzki and S. Zorn, *Financing Mining Projects in Developing Countries* (London, Mining Journal Books Ltd., 1979).

30. S. Koutoubi and L.W. Koch, "Uranium in Niger," in *Uranium and Nuclear Energy* (London, Mining Journal Books Ltd., 1980).

Observation of these enterprises suggests that, at least in their early lives, their owner governments have abstained from exerting any strong influence over their activities. The arrangements with foreign parties usually include the clear understanding that the government has no managerial and technical experience with the activity and that the foreign joint venturers are to take full responsibility both for the development and for subsequent operations. The purpose is to ensure an early revenue stream. Hence, the behavior of these enterprises is much like that of ordinary private enterprises. With time, as the government administrations learn to run mineral enterprises, they can be expected to influence an increasing number of functions and to change the goals and operating modes of the mineral activity toward the patterns typical of the experienced state-owned mineral firm. The shift would ordinarily be gradual. Thus, the emancipation of this second type of inexperienced state mineral firm in developing countries is likely to cause far less disruption in marketing, production, and investment activity and hence involve smaller start-up costs than in the case of the first type of state-owned mineral company.

Market Impact of State Mineral Enterprises

The analysis presented so far suggests a number of plausible suppositions about the characteristics and behavioral patterns of state-owned mineral enterprises in developing countries and about some of the market implications of these findings. The analysis will now be carried one step further to formulate in a more systematic way a set of hypotheses about how the increasing importance of state enterprises as mineral suppliers affects international mineral markets. Each of the hypotheses differentiates between market situations in which state enterprises account for a significant share of supply and those in which private firms are completely dominant. Each of the hypotheses is justified by a restatement of the relevant findings and suppositions from the earlier analysis.

a. The widespread nationalizations of mineral activities in developing countries during recent decades have temporarily reduced output and raised prices compared to what would have prevailed under an uninterrupted private multinational regime. Takeovers by the state have regularly involved substantial start-up and learning costs in the form of reduced efficiency, increased cost levels, and inability to operate existing installations at full capacity or to establish new capacity. The logical consequence of this development is clear. Because the nationalizations were usually sudden events and in many cases unexpected by the industry,[31] and given the long gestation period of new investments in mineral production, private industry could not immediately compensate for the shortfall in supply after the state enterprises were set up. Hence, until private firms could increase output sufficiently or the nationalized firms could acquire enough expertise to invest in capacity expansion, prices increased more than they would have in the absence of the nationalizations. This market impact should be of a transient nature and should diminish as the state-owned firms gain experience. The transition process is gradual, however, and may take ten to twenty years in any individual case.

b. The takeover of mineral production by state enterprises will permanently raise production costs because mature state firms incur additional costs in their pursuit of nonprofit objectives and because they are generally under less pressure to minimize costs. These higher costs will normally absorb part of the mineral rent in the intramarginal firms. Where the state operates marginal ventures, they will require permanent subsidies or else will be forced out of business.

c. The increasing importance of state enterprise will result in greater medium-run price instability in most international mineral markets because state mineral firms are less flexible in adjusting capacity utilization to cyclical variations in demand.

d. Nationalizations in developing countries have frequently broken up multinational chains of vertical integration. The national character of state-owned firms makes them reluctant to reestablish vertical integration through direct downstream investments abroad, and the extent of economical vertical integration at the national level is limited. Hence, the entry of state enterprises will permanently reduce the degree of vertical integration in the overall industry and thus broaden the arms-length markets for mineral products, especially at early production stages. In such industries as iron ore and bauxite, the establishment of state-owned firms has significantly expanded the formerly very narrow noncaptive markets, thereby contributing to price stability in the noncaptive trade.

e. The preceding analyses do not lend support to the thesis that the emergence of state-owned firms in developing countries threatens the survival of the private mineral industry. The private sector's share of world markets has

31. R. Vernon, *Two Hungry Giants*.

shrunk with the establishment of state enterprises and may shrink even more if and when further nationalizations in developing countries occur. The analysis of this chapter suggests, however, that state enterprises' operations and investments are less cost efficient than those of the private sector and that permanent subsidies to the state-owned sector are unlikely. Further, neither the empirical observations nor the logical deductions presented and explored in this chapter indicate a greater investment propensity in state-owned mineral enterprises than in private ones. It follows that a case for threat to survival of private firms cannot be maintained.

II

THE CASE STUDIES

Pertinent facts about the companies treated in the three case studies are listed in the table below. Timah is the oldest under government ownership. In 1980, it was by far the most profitable one, both in relation to total capital and to sales. ZCCM is much larger than the two others. Its capital, turnover, and employment are several times the size of Timah's and Ferrominera's. The Venezuelan iron ore company is the most capital intensive of the three. In 1980, its capital assets per employee were worth nearly $95,000 (U.S.).

ZCCM stands out among the three companies in that it dominates the Zambian economy in several respects. This company's sales in 1980 corresponded to one-third of Zambia's gross domestic product and accounted for more than 95 percent of the country's export revenues. In contrast, only 2 percent of Zambia's working-age population was employed in this capital-intensive mineral-mining and -processing corporation. The corresponding measures for the other two companies show that Timah and Ferrominera are relatively insignificant in all three respects in their national economies.

All three companies are very heavily oriented toward the international market, both in their sales and in their purchases of capital equipment. For this reason, but also because the present study is concerned with international markets, all values relating to the companies are presented in U.S. dollars, obtained by converting the national currencies at actual exchange rates. The use of an international currency creates anomalies in those series of values that are mainly nationally based. A devaluation of the national currency will tend to produce a discrete reduction of the cost figures in U.S. dollars or of the owner capital. These anomalies would usually be of limited duration, how-

The Three State Mineral Companies Compared, 1980

Item	PT Timah, Indonesia	Ferrominera, Venezuela	ZCCM, Zambia
Year state ownership began	1950	1974	1970
Total capital (m$)	414[a]	419	1,850
Sales (m$)	410	312	1,312
Pretax profits (m$)	192	a	284
Employment, number of persons	27,650	4,440	57,750
Capital per employee ($)	14,980	94,370	32,030
Sales as a percentage of GDP	0.59	0.52	36.60
Sales as a percentage of exports	1.87	1.52	96.70
Employment as a percentage of working age population (15–64 years)	0.03	0.05	1.99
Production as a percentage of world output	11.1	1.8	7.8

Sources: Chapters 4, 5, and 6 of this volume; World Bank, *World Development Report*, 1982; Bank of Zambia, *Annual Report*, 1981.

[a] 1978.

ever, because devaluations are typically preceded or followed, or both, by higher rates of inflation in the currency that is devalued. In any case, the use of an international currency is necessary for comparisons both among the three companies studied and in relation to their international competitors.

Where relevant, current-dollar values are converted into constant (1981) dollars. The World Bank's Manufactured Exports Unit-Value Index CIF (cost, insurance, freight), expressed in U.S. dollars (formerly known as the World Bank Index of International Inflation) is used as the deflator to obtain the constant-dollar series. This index is used in preference to a domestic U.S. series, such as the U.S. producer price index, because it gives a better measure of the international value of the dollar. For instance, between 1980 and 1982, the U.S. producer price index increased by 11 percent, but because the dollar appreciated sharply against most other currencies during the period, its international purchasing power rose by 4.5 percent, despite the domestic U.S. inflation. Because the contacts of the companies studied are international, it is the international purchasing power of the dollar that is relevant.[1] The World

1. Although the deflator chosen appears appropriate for the purposes of the present study, it is not very meaningful for some other objectives. For example, a time series of unit labor costs expressed in constant dollars gives one measure of the international competitiveness of the company studied. But this series does not say much about changes in welfare and consumption standards of the labor force. Local currency and local price indexes would be needed to obtain the latter.

Bank's index for 1961 to 1982 is shown in the table below.[2]

The World Bank's Manufactured Exports Unit-Value Index, CIF
(in U.S. dollars, 1981 = 100)

Year	Index	Year	Index	Year	Index	Year	Index	Year	Index
1961	28.9	1966	31.2	1971	35.8	1976	68.0	1981	100.0
1962	28.6	1967	31.7	1972	39.3	1977	73.7	1982	100.5
1963	28.8	1968	29.7	1973	46.9	1978	86.9		
1964	29.4	1969	29.9	1974	58.8	1979	97.0		
1965	29.5	1970	33.1	1975	66.8	1980	105.0		

The metric ton is almost exclusively used as the weight measure in this study. The only exception is that, following the predominant practice of the mineral trade literature, unit prices and costs of tin and copper are given per pound (454 grams).

The three case studies have, by and large, the same format. They all begin by describing the history of the particular mineral industry, and the circumstances that led to its nationalization. The goals of each state corporation are identified, and its achievements are then explored. Each study ends by pointing to the distinguishing features that have followed from state ownership.

2. World Bank, *Commodity and Price Trends*, 1982/83 edition, updated and revised according to information received from the World Bank in April 1983.

4

The State Tin Industry in Indonesia

The Dutch initiated tin mining in Indonesia in the early part of the 18th century.[1] In the years immediately following the second world war, the colonial government of the Netherlands East Indies was the sole owner of the largest production center and held majority equity positions in the remaining two production units. All three were operated under long-run management contracts by Billiton, a private Dutch company named after one of the tin-producing islands of Indonesia. Billiton also had some equity in the two smaller units. In 1950, at independence, the national government took over the equity positions held by the colonial administration. In 1958, the government also acquired Billiton's Indonesian equity. From that year until 1973, when Koba Tin started production, the ownership and management of Indonesia's tin industry remained completely in the national government's hands.[2]

Initially, the three existing production units were operated as separate companies. In 1961, the government established the General Managing Committee of the State Mining Companies, which acted as a unifying umbrella for the tin industry. In 1968, all production operations were merged into a single state

Note: The data upon which the analysis is based have been obtained from material published in Indonesia and elsewhere, as quoted, and, above all from a series of personal interviews with officials of the ministries of Mines and Energy and Finance, with managers of the tin enterprises and others conducted during a two-week visit to Jakarta in October 1981. A list of my interviews is given in Appendix 4.A.

1. *Tin in Indonesia*, undated brochure published by the Public Relations Division of PT Timah, the state tin corporation, circa 1977.

2. W. Fox. *Tin, The Working of a Commodity Agreement* (London, Mining Journal Books Ltd., 1974), and interviews with retired officials of the Ministry of Mines.

enterprise, Perusahaan Negara (PN), under the name of PN Tambang Timah. This firm, almost exclusively, carried out the whole range of activities of the tin industry, from exploration, mining, processing, and smelting through to final marketing. Finally, in 1976, the company was reorganized into a limited liability company, Perusahaan Terbatas (PT), 100 percent owned by the government, with considerably expanded commercial independence.[3] The present state-owned company, PT Tambang Timah, as well as its predecessor PN Tambang Timah, will be referred to in what follows simply as Timah.

In 1967, the incoming administration of President Soeharto introduced an important policy change with regard to the tin industry: Indonesia's mining sector was opened to foreign investments.[4] Three new tin ventures with foreign involvements have been established as a result. PT Koba Tin, 75 percent owned by Australian interests, with a 25 percent equity holding by Timah, started its exploration program in 1971 and went into production in 1973. In 1980, this company's production reached 5,260 tons of tin in concentrates, 16 percent of the national total. In that year, Koba Tin had about 2,200 employees on its payroll, of whom 24 were expatriates.[5] The entire output of concentrates was sold to Timah, which smelted and marketed it, along with its own produce.

PT Riau Tin Mining, the second foreign tin venture, was initiated by Billiton of the Netherlands, since 1970 a subsidiary of the Royal Dutch Shell Group. Timah holds 10 percent of the equity in Riau Tin and there is a commitment to offer a further 15 percent to the Indonesian public. Thus, like Koba Tin, the final foreign equity involvement will amount to 75 percent. Production started on a small scale in 1979, and the 1980 output amounted to 630 tons of tin in concentrates. The level of production is expected to rise to 1,000 tons, as full capacity is reached in the early 1980s. In 1980, Riau Tin had 320 employees. The concentrate output was smelted by Timah on a toll basis, and the tin metal was subsequently marketed by Riau Tin itself.[6]

PT Broken Hill Proprietory Indonesia (BHP Indonesia) is the third foreign venture, owned 100 percent by Australian interests. The property of this company consists of a rehabilitated underground mine abandoned in 1941 at the time of the Japanese invasion. Production operations started in 1975. By

3. PT Timah, *Tin in Indonesia*.

4. S. Sigit, Ministry of Mines and Energy, "Mineral Resources for the 21st Century, Challenges and Opportunities, an Indonesian Viewpoint," contribution to the U.S. Geological Survey International Centennial Symposium, October 1979.

5. *Indonesia Development News*, no. 7, 1981.

6. Information from interview with General Manager of Riau Tin.

Table 4-1. Indonesian and World Mine Production of Tin

Item	1950	1955	1960	1965	1970	1975	1980	1981	1982
Indonesia (000t)	32.6	33.9	23.0	14.9	19.1	25.3	32.5	35.3	33.8
World (000t)	176.9	194.0	188.8	203.7	217.1	218.1	235.0	237.1	188.1
Indonesia, share of world (%)	18	17	12	7	9	12	14	15	18

Source: Metal Statistics, several issues.

Table 4-2. Indonesian Production of Tin, by Company

Company	1975	1976	1977	1978	1979	1980	1981	1982
Riau Tin (tons)					95	630	1,032	}
BHP Indonesia (tons)	79	210	296	433	469	500	525	} 7,581
Koba Tin (tons)	867	1,022	1,610	2,914	3,807	5,260	6,532	}
Timah (tons)	24,391	22,203	24,021	24,064	25,164	26,110	27,179	26,219
Total (tons)	25,337	23,435	25,927	27,411	29,535	32,500	35,268	33,800
Share of Timah (%)	96	95	93	88	85	80	77	78

Sources: Indonesian Mining Yearbook 1979, for 1975 through 1979; company information; *Metal Statistics*, 1983, for 1980 through 1982.

1980, output had reached the level of 520 tons of tin in concentrates. Full capacity of about 1,000 tons was expected to be attained in the early 1980s after completion of further investments. In 1980, BHP Indonesia employed 450 persons, of whom some 20 were expatriates. The company itself is responsible for marketing its output internationally after toll smelting in the UK and Malaysia.[7] Smelting takes place outside Indonesia because the particular chemical composition of the concentrates makes them less suitable as feed in the Timah smelter.[8]

Table 4-1 traces the development of Indonesia's tin output since 1950 and compares it with the world total. After reconstruction in the immediate postwar period, Indonesia reached its production peak of 36,400 tons (18 percent of world output) in 1954. Production fell continuously during the following twelve years, to reach a minimum of 12,800 tons (6 percent of world output) in 1966. Since then, production has been rising steadily. In 1982, output reached 33,800 tons, just above the level of 1950, and again representing 18 percent of the world total. Table 4-2 details the different companies' production from 1975 to 1982.

Timah's share of the total declined from 100 percent in the early 1970s to 96 percent in 1975 and 78 percent in 1982, as the foreign ventures attained their

7. Information from interview with Public Relations Department of BHP Indonesia.

8. Information from interview with Timah personnel.

full capacity output. Because no further foreign ventures are planned and because private Indonesian enterprise is prohibited by law from entering the tin industry,[9] Timah will continue to dominate Indonesia's tin, both in share of mine output and in its complete control of domestic tin smelting.

State Enterprise—The Early Period

The nationalization of the tin industry followed a somewhat unusual pattern. Prior to independence, majority ownership was not in private foreign hands but was held by the colonial government. At independence, transfer of the colonial power's equity positions did not involve a takeover by purchase or confiscation. The assets were simply transfered to the national administration, in much the same way as, for instance, the railways, the postal service system, and other facilities run by the public colonial authorities. In 1958, the government also agreed with Billiton to swap the private company's remaining tin involvements in the country for all the country's tin-related corporate assets outside Indonesia.[10]

The existing management contracts held by Billiton were allowed to continue until their expiration—in 1953 for one production center and in 1958 for the other two. The animosity resulting from the independence struggle led the Indonesian government to decide to completely nationalize management functions when the contracts with Billiton expired, despite the paucity of Indonesians with technical and administrative qualifications.

The period from the latter half of the 1950s and until 1966 was a difficult one, not only for the tin industry, but for Indonesia's economy as a whole. The task of nation building, involving unification of a geographically dispersed area with wide ethnic and cultural differences, took priority over economic management.

The fall in tin production, from 36,400 tons in 1954 to 12,700 tons in 1966, or from 18 percent to 6 percent of world output, has two major explanations. The first is the inexperience of the national management cadre that took over when the Dutch left. In the Dutch management period, all higher level positions in the tin companies were held by non-Indonesians. The first Indonesian engineering and geology graduates, emerging from the universities only in the late 1950s, were put in charge of the installations, despite their complete lack

9. Information from Timah.

10. Interview with Mr. M. Subroto, former Director General in the Ministry of Mines and Energy.

of practical experience. Undoubtedly, this must have had a severe negative impact on the efficiency of operations. The inattention to planning and maintenance of installations was particularly detrimental to long-run output trends. The managerial inadequacies continued long beyond the Soekarno era. In the view of one seasoned Indonesian observer, not until the early 1970s did the national management teams acquire the technical, administrative, and commercial proficiencies of an international standard and overcome the deficiencies caused by managerial neglect in earlier years.[11] The problems caused by lack of managerial experience could certainly have been overcome in some measure by reinforcing the national teams with foreign expertise. In the strongly nationalistic mood that characterized the country at that time, however, such measures were considered politically inopportune.

The second reason for the drastic reduction in output was that the government, in a general effort to maximize the resources immediately available for the political task of nation building, drained the industry of funds. Politics came first during that period, and economic and social development were secondary issues. Most of the state tin companies' records from this period, if they exist at all, are nonaccessible. Hence, precisely how the government treated the industry cannot be documented. At times, only one-quarter of tin export proceeds were credited to the producing enterprises, and these credits were calculated in a strongly overvalued local currency.[12] The consequences for the tin industry were highly detrimental. The government appropriated such a large share of the net cash flow of the tin mining enterprises that little was left even for maintaining existing facilities, much less for financing exploration to replace depleted deposits with new ones or for investing in expanded production capacity.[13] A number of high-cost units had to be closed down, and their equipment cannibalized, to limit the financial drain and so permit continued operation of the remaining installations.

This period saw only one major new investment venture in Indonesia's tin industry. In that case too, the decision was based on political rather than economic grounds. In an effort to reduce its dependence on the Netherlands, where most of Indonesia's tin concentrate was sent for smelting, the government decided in 1959 to establish a tin smelter. The political appropriateness of this decision was confirmed by the subsequent conflict with the Netherlands about Irian Jaya, which prompted Indonesia to shift concentrate exports

11. Interview with Soetaryo Sigit, Secretary General, Ministry of Mines and Energy.

12. Interview with Mr. M. Subroto.

13. M. Gillis and R.E. Beals, *Tax and Investment Policies for Hard Minerals: Public and Multinational Enterprises in Indonesia* (Cambridge, Mass., Ballinger Press, 1980).

toward the United States and Malaysia. The conflict with Malaysia in 1963 forced a new redirection of exports, from Malaysia back to the Netherlands.[14]

In 1961, Indonesia signed a contract with Klockner Industries of West Germany to install a smelter of a new design with a capacity of 25,000 tons of metal a year, at a cost of $3.6 million. Most of this capital was provided as supplier credit. Construction was completed in 1967. Technically, the venture must be considered a failure. By 1970, after three years of trial runs, the annual output did not exceed 5,000 tons. After considerable technical alteration, output reached about 15,000 tons in 1973.[15] The relative inexperience of the technical personnel in the Indonesian tin industry at that time undoubtedly contributed to the failure, but similar difficulties occurring in Bolivia, where a Klockner tin smelter using the same technology was established at about the same time, indicates the probability of serious problems with Klockner's technical design, as well.[16]

Ironically, the smelter proved profitable to Timah, despite its technical difficulties. This was because the government itself underwrote the obligations for supplier credit and simultaneously extended a low-interest, local-currency loan to the company to enable it to finance the investment. Ensuing devaluations virtually wiped out the value of this loan. Most of the cost of the investment was therefore incurred by the government.[17]

The general patterns and rules for the operations of the tin industry during this traumatic period are difficult to establish. Quite clearly, the government intervened very heavily in the operations of the tin enterprises on an ad hoc basis. Production of tin was seen as a tool for promoting the government's political goals. Little consideration was given to the long-run consequences for the industry following from the interventions.

State Enterprise Since 1967

The year of 1967, when President Soeharto established his "New Order" government, constitutes a clear-cut dividing line in Indonesia's state involvement in tin. The new administration introduced very profound changes in government policy towards the tin sector.

14. W. Fox., *Tin, The Working of a Commodity Agreement.*

15. K.A. Batubara, "Tin Smelting in Indonesia." Paper presented at the Indonesian Mining Association Symposium in Jakarta, June 1977.

16. Private communication with Malcolm Gillis.

17. Interview with Mr. M. Subroto.

The activities of Timah are far easier to record and quantify in the period from 1967 and onwards. Nevertheless, data on Timah have proved more difficult to extract than those on the state-owned mineral companies in Venezuela and Zambia. In October 1981, during my visit to Jakarta, the 1978 annual report was the most recent available for Timah. Even in August 1983, when Timah responded to my request for additional data, it did not supply more recent annual reports.

Direction and Goals

Subject to the current order of Indonesia's president, the ministers of mines and energy and of finance share oversight responsibilities for Timah on behalf of the shareholder (government). After hearing recommendations, the president appoints the Board of Commissioners (three in number), which supervises the activities of the enterprise, and the Board of Directors (one president and four members, each responsible for a technical function), which manages Timah on a day-to-day basis. The government's formal control responsibility includes the scrutiny and approval, both by the Board of Commissioners and by the two ministers, of the annual budgets the Board of Directors prepares. Timah's accounts, financial dispositions, and investment expenditures are also scrutinized by special public bodies set up to control all public enterprises in these respects.

Possibly as a result of the devastating experiences of excessive government involvement in the earlier years, the administration has by and large abstained during this later period from direct intervention in the operations of Timah. For instance, the Board of Directors' annual budget proposals on investments, production, and disposition of funds are usually accepted without amendments by the two responsible ministers.

Neither the government nor the company itself has ever clearly defined the objectives or outlined the long-term strategy the publicly owned tin industry is to follow. Discussions with officials in the Ministry of Mines and Energy and with the managers of Timah suggest that the firm pursues long-run profit maximization, subject to the fulfillment of a variety of far-reaching social obligations, self-imposed by the management or specially decreed by the government. As an example of the latter, the government has required the tin corporation to provide employment to groups of retired army officers. Expansion of the company's activities appears to have been determined by short-term considerations on an ad hoc basis.

In a reorganization undertaken in 1976, the profit orientation of Timah was given more emphasis and the company gained greater freedom to pursue profitable activities, even outside the tin industry. Also, beginning with that year, incentives have been provided to the management and employees for improving the company's profit performance. In 1976 and 1977, for example, 20 percent and 15 percent, respectively, of aftertax profits were spent on management bonuses and on wage increases for other employees. (Prior to 1976, bonuses to management and employees were considered a cost unrelated to profits.)

Despite these developments, the management expresses the view that pressures of public accountability restrict corporate flexibility and slow down decision making. For fear of breaking some bureaucratic rule in matters pertaining—for instance, to employment or company purchases—managers must think twice before acting and often prefer not to act at all, even when an action clearly would benefit the company. This, in the opinion of the managers, results in opportunity losses and makes managerial tasks more unwieldy than in private Indonesian enterprises. Timah's internal organization is more centralized than is common in multinational corporations of corresponding size. All important decisions are referred to the directors in charge, and the department heads have limited freedom for independent action. This feature probably stems partly from the geographical concentration of Timah's operations, but state ownership is certainly another contributing cause. Thus, by blocking action and centralizing decision making, state ownership adds to the inflexibility of the enterprise.

The Tin Agreement

The prices received by tin producers throughout the world are based on the Penang market or the London Metal Exchange (LME) quotation.[18] The two follow each other reasonably closely. Figure 4-1 depicts real price developments for tin since 1960. Despite slow growth in world consumption (less than 1 percent a year on average during the 1950–82 period), the price trend has been quite favorable from the producers' point of view, both in itself and in comparison with most other raw materials. For instance, while the index of nominal dollar prices for tin (1960–64 = 100) reached 400 in the 1976–80 period, that for all minerals, ores, and metals (fuel minerals not included) reached only 260.[19] The tin producers certainly had more favorable price

18. W. Labys, *Market Structure, Bargaining Power and Resource Price Formation* (Lexington, Mass., Lexington Books, 1980) p. 125.

19. United Nations Conference on Trade and Development, *Monthly Commodity Price Bulletin, 1960–1980 Supplement*, April 1981.

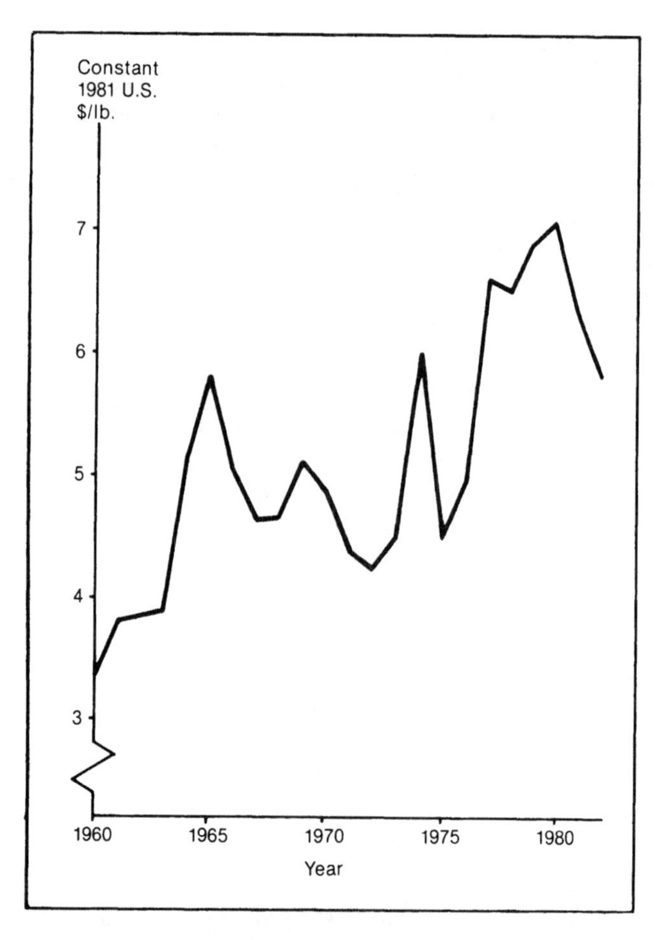

Figure 4-1. Average Annual Tin Prices, Penang Market, 1960–82 ***Source*****: UNCTAD,** ***Monthly Commodity Price Bulletin*****, various issues.**

tendencies during this period than the producers of iron ore and copper (see chapters 5 and 6).

Price developments in the short and medium run have been influenced by the operations of the International Tin Agreement and by disposals from the U.S. strategic stockpile. The major purpose of the tin agreement, established in 1956, has been to maintain prices within a range determined at irregular intervals by the producing and consuming members of the International Tin Council (ITC). As a prominent member of the Council, Indonesia has had to conform to the provisions imposed by the agreement. The instruments for price control consist of a buffer stock and quotas on exports, combined with limitations on permitted stocks in producing countries. Since 1960, the export quotas were so mild that they did not require production cuts in the exporting countries.[20] The tin agreement has been successful in defending the floor prices—since 1958, the market price has never broken through the floor.[21] On the other hand, the price has risen above the ceiling on several occasions.

The major role played by releases from the huge U.S. strategic stockpile (190,000 tons in 1983, or equal to almost one year's world consumption) has been to subdue the price explosions that the tin agreement could not control. Simulations carried out by Smith and Schink suggest that these releases have been more powerful than the tin agreement programs in reducing medium-term price fluctuations.[22]

Timah's Record

Table 4-3 provides production and export figures for Timah from 1967 to 1982. In the first eight years of President Soeharto's "New Order," the company's output almost doubled, as worn-down installations were rehabilitated and new facilities were established. After 1974, however, output varied from year to year but without any upward trend. No more than 5 percent of Timah's production is domestically consumed. The International Tin Agreement has prevented Indonesia from exporting freely at all times. Although, technically, capacity utilization has been maintained through the period covered by the table, part of the output has been added to inventories in years when the Council imposed export restrictions. Exports exceeded mine production in the late 1970s because Timah was marketing increasing amounts of tin mined by Koba Tin.

In distinction from the producers in Malaysia, Thailand, and Bolivia, who sell most of their output either directly in Penang or through dealers, Timah has developed a sophisticated system of marketing most of its tin directly to the final consumers.[23] The contractual arrangements with customers are oriented towards long-run profit maximization and subordinate the short-run benefits of such objectives as long-run stability and diversification of sales.

Table 4-4 provides a rough outline of the changes in Timah's capital structure in the ten-year period, 1969 to 1978. Although the company's debt-equity

20. G.W. Smith and G.R. Schink, "The International Tin Agreement: A Reassessment," *Economic Journal*, December 1976.

21. World Bank, "Price Prospects for Major Primary Commodities," Report No. 814/82, July 1982.

22. Smith and Schink, "The International Tin Agreement."

23. W. Labys, *Market Structure*, p. 125; and information from interview with Timah.

Table 4-3. Timah, Mine Production and Exports of Tin 1967–82
(tons)

Year	Mine production	Exports
1967	13,827	10,551
1968	16,939	17,960
1969	17,416	17,210
1970	19,091	17,588
1971	19,765	18,874
1972	21,765	20,603
1973	22,491	20,969
1974	25,023	22,983
1975	24,391	22,011
1976	22,203	24,111
1977	24,021	24,914
1978	24,064	25,549
1979	25,164	25,736
1980	26,110	30,785
1981	27,179	n.a.
1982	26,219	n.a.

Sources: 1966–75, *Tin in Indonesia*; 1976–82, company information.

ratio has risen somewhat, its solidity has remained quite strong, with loans never constituting more than 30 percent of total funds. Most of the loans are of short duration; the company's long-term debt is insignificant.

Socially, Timah has been a highly profitable operation through the 1969-78 period, for which full data are available. An export levy amounting to about 10 percent of sales revenue has been added to cost, and the profit that emerged thereafter has been subject to the company income tax at about 45 percent. The aftertax profit, $198 million in all, corresponded to an average return of 10.7 percent on owners' capital. About 20 percent of aftertax profit has gone to shareholder dividends. What remained was more than adequate to finance gross investments, even without making use of the cash flow from depreciation. The amounts of total depreciation in the years under scrutiny have been unobtainable.

The financial returns to Indonesia from Timah's activities should rightly include the profit before tax plus the export levy. As may be computed from table 4-5, these amounts total almost $500 million in the ten years and equal an average return of just above 28 percent on shareholders' funds.

The favorable social return realized by Timah does not necessarily imply that the company was run in a cost-efficient way. A major proportion of Indonesia's tin output is extracted from some of the world's most economical

Table 4-4. Timah, Debt and Own Capital, 1969–78
(million $)

Item	1969	1970	1971	1972	1973	1974	1975	1976	1977	1978[a]
Short-term debt								66	72	122
	20	22	34	44	45	52	89			
Long-term debt								12	11	7
Own capital	112	116	108	110	119	134	140	170	227	285
Total funds used	132	138	142	154	164	186	229	248	310	414

Source: PT Timah, Annual Reports.

[a] On Nov. 15, 1978, Indonesia's currency was devalued by 50 percent, but the annual report from which the figures are taken provides the dollar amounts calculated on the basis of predevaluation exchange rates.

Table 4-5. Timah, Summary Profit and Loss Account, 1969–80
(million $)

Item	1969	1970	1971	1972	1973	1974	1975	1976	1977	1978[a]	1979	1980[b]
Sales proceeds	62.7	65.3	65.3	72.8	100.6	174.1	166.6	188.3	249.6	353.2	398.4	410
Total cost	50.2	49.4	47.7	63.0	84.0	148.6	150.7	151.4	178.5	225.5	227.4	218
Production and marketing	41.5	41.6	n.a.	54.5	66.7	105.3	109.6	132.9	137.7	158.4	—	—
Export tax	8.2	6.7	5.9	7.1	8.8	17.2	17.1	16.0	24.3	33.8	—	—
Profit before income tax	12.5	15.9	17.6	9.8	16.6	25.5	15.9	36.9	71.1	127.7	171.0	192
Income tax	0.7	6.8	7.1	4.0	8.4	13.3	8.3	17.7	31.6	53.4	—	—
Profit after income tax	11.8	9.1	10.5	5.8	8.2	12.2	7.6	19.2	39.5	74.3	—	—

Sources: 1969–76 from M. Gillis and R. Beals 1980; 1977–78 from Timah *Annual Report 1978*; 1979–80 statistics released by Timah in February 1981, as reported by the Embassy of Canada in Jakarta.

[a] On Nov. 15, 1978, Indonesia's currency was devalued by 50 percent, but the annual report from which the figures are taken provides the dollar amounts calculated on the basis of predevaluation exchange rates.

[b] Preliminary figures.

deposits,[24] and this alone should enable the operations to reap substantial mineral rents. To determine Timah's cost efficiency, then, requires probing deeper into the company's activities.

Before venturing into that exercise, a comparison of Indonesian (national) cost levels with those of other tin-producing countries may be instructive. Table 4-6, mainly based on data released by the International Tin Council, provides such a comparison. The table basically contains three series of figures, all expressed in U.S. cents per pound of tin. The first one is the average annual tin price in the Penang market from 1969 to 1981. The second compares the production costs in Indonesia with an average for all producing members of ITC, as reported to the Council by each country. Note that beginning in 1973, the Indonesian figure is a weighted average of Timah and the foreign operations, although the state company dominates the total throughout the period. The third series provides the average cost of producing and marketing tin in Timah, recalculated from table 4-5, and originally obtained from this company's accounts. The differences between this cost series and that reported to the ITC are unclear. Timah's figures probably do not take full account of the costs of depreciation and on-site exploration, which are included in the ITC numbers. All figures are expressed in current as well as in constant (1981) dollars.

Table 4-6 shows that until 1979 Indonesia's production cost was above the average for all producing ITC members. A comparison of Indonesian cost levels with those in Malaysia, the largest tin producer, is revealing. Malaysian average costs have persistently been lower than those of Indonesia, despite the fact that Malaysia's mines are much deeper and the ore being exploited are of much lower grade.[25] In the late 1970s, the wages of Malaysian tin workers were almost twice as high as in Indonesia.[26] These circumstances strongly suggest that Timah, the dominant Indonesian producer, is not as cost efficient in its production as its Malaysian competitors. Cost levels in Thailand also were lower than those in Indonesia through the period studied.[27] Hence, the ability of Timah to generate economic rents per unit of tin output has been dwarfed by that of the Malaysian and Thai tin producers.

The response of Timah's management to the allegation of cost inefficiency is that the elevated Indonesian cost level arises primarily from the company's

24. Interview with Soetario Sigit, Secretary General, Ministry of Mines and Energy.

25. J. Thoburn, *Multinationals, Mining and Development, A Study of the Tin Industry*, chapters 5 and 6 (Brookfield, Vt., Grover Publishing Co., 1981).

26. Ibid.

27. ITC statistics obtained from sources quoted in table 4-6.

Table 4-6. Tin Prices and Average Production Costs, 1969–81
(U.S. cents/lb.)

Item	1969	1970	1971	1972	1973	1974	1975	1976	1977	1978	1979	1980	1981
Current dollars:													
Tin price, Penang[a]	153	162	157	168	212	354	302	338	485	567	672	744	637
Costs excluding royalty and export tax, as reported to ITC[b]													
Indonesia	—	—	132	128	154	251	227	249	330	370	326	424	479
Weighted average for producing members of ITC	—	—	114	120	138	195	219	229	283	334	370	456	468
Timah, production and marketing costs[c]	90	90	108	114	135	191	196	258	244	262	—	—	—
1981 dollars:													
Tin price, Penang[a]	512	489	439	427	452	602	452	497	658	652	692	709	637
Costs excluding royalty and export tax, as reported to ITC[b]													
Indonesia	—	—	369	326	328	427	340	366	448	426	336	404	479
Weighted average for producing members of ITC	—	—	318	305	294	332	328	337	384	384	381	434	468
Timah, production and marketing costs[c]	301	272	302	290	288	325	293	379	331	302	—	—	—

Sources: Price—[a]UNCTAD, *Monthly Commodity Price Bulletin 1960–1982 Supplement*, April 1983. Costs—[b]1971–1975, from W. Labys, *Market Structure Bargaining Power and Resource Price Formation*, Lexington Books 1980, p. 113; 1976–1981, from data prepared by International Tin Council, obtained from Timah. Timah—[c]1969–1977, from M. Gillis and R.E. Beals, *Tax and Investment Policies*, p. 82; 1978, from Timah *Annual Report*.

social obligations, voluntary or forced, that do not apply generally to other enterprises or to the other tin producers in Indonesia. These social obligations pertain mainly to Timah's labor force, to community development, and to regional policy.

Total employment in the company rose from 25,000 in 1966 to 27,000 in 1968[28] and has varied between 27,000 and 28,000 up to 1980. Output per employee increased from 0.5 tons in 1966, to 0.9 tons in 1975, and to almost 1 ton in 1980. Even in 1980, however, Timah's labor productivity was much lower than in the private tin operations in Indonesia. The management of the state corporation readily admits that the company has more employees than it needs. For instance, of a labor force of 27,652 persons in December 1975, 675 were classed as "honorary personnel" and 3,904 as "excessive."[29] Although wages are lower than those offered for similar jobs in the private sector, the workers benefit from a variety of costly welfare programs, far exceeding standard custom in the country. Downward adjustments in the labor force are hampered by rules for job security, which in turn result in an extremely low labor turnover of about 2 percent a year. In 1980, labor costs constituted about 20 percent of total production and marketing costs.[30] The combination of overstaffing and low wage levels in Indonesia as compared to Malaysia and Thailand result in very similar labor cost shares in total tin production costs in the three countries.[31]

Timah feels responsible for providing community development services in the areas surrounding its operations, even though it has no legal obligation to do so. For example, the company supplements the local public authorities in financing housing projects, in supplying medical and educational services, in repairing roads, and in subsidizing household electricity. The company supports the national government's regional policies by maintaining unprofitable operations in areas with little alternative economic potential and by considering regional development issues along with expected profitability in ranking potential investment projects. These social obligations are costly to Timah. In the management's view, the company's costs might be reduced by about 15 percent, if it did not accept social obligations beyond those mandatory for a private corporation operating in the country.[32]

28. PT Timah, *Tin in Indonesia.*

29. Ibid.

30. Interviews with management of Timah.

31. J. Thoburn, *Multinationals Mining and Development.*

32. Interview with management of Timah.

Timah contracts with local entrepreneurs to mine small deposits located away from the major production centers. Some 10 to 15 percent of total output originates from such operations. The management rationale for using private contractors is that such an arrangement simplifies and reduces the need for control in Timah. The use of private contractors can also be seen as a measure to avoid some of the social costs of in-house operations and as a means of increasing the flexibility of overall operations, for instance, in labor mobility or production techniques. An additional reason may be to take advantage of Chinese mining expertise, because company statutes prevent Timah from employing ethnic Chinese directly.[33]

Managerial inexperience and lack of technical competence are no longer likely to contribute to Timah's elevated cost levels, but a lax approach to cost control, along with the expensive social obligations, could explain why—despite its advantageous natural resource position—the company's record compares unfavorably with those of the tin producers of Malaysia and Thailand. One reason for the easy-going attitude toward cost minimization may be the very considerable job security Timah's managers enjoy. All of the half-dozen managers I interviewed had been with the company for twenty years or more. Another likely reason might be that, because the company generates such high profits, the government has not insisted on maximum feasible cost reductions. Finally, Timah's unclearly defined multi-purpose objective makes cost control more difficult.

Table 4-6 permits scrutiny of the development of costs over time, as distinct from the cost level at any one time. In the 1971–78 period, production costs in Indonesia, as reported to the ITC, rose by no more than 15 percent in real terms, as compared to a 21 percent increase for the international average. Timah's costs, as reported in its own accounts, do not appear to have risen at all in that period. This above-average cost containment is explained in considerable measure by a large-scale shift from on-shore to off-shore production during the 1970s. An additional explanation could be that, at least in the early 1970s, the company was still recovering ground lost in the period after national takeover.

Cost developments in 1979 and 1980 have been exceptional. In 1979, Indonesian cost levels, as reported to the ITC, fell in absolute terms and remained below the international average into 1980. Two factors outside Timah's control explain this favorable development. First, the Indonesian currency was devalued by 50 percent in November 1978, and second, domestic fuel prices were not permitted to rise in 1979 and 1980 in spite of the sharp increase in

33. Private communication with John Thoburn.

international petroleum prices. In these two years, therefore, Timah's international competitiveness was significantly improved by an increase of the implicit fuel subsidies. In 1980, the company's costs might have been 7 to 8 percent higher, if it had paid the international price for the fuel it used.[34] By 1981, however, the advantages of devaluation had been eaten up by accelerated inflation, and Indonesian tin-producing costs again exceeded the ITC average.

Has Timah ever deviated from profit-maximizing behavior in the sense that its marginal cost exceeded price as a concomitant of full capacity utilization? The cost data do not permit a definite resolution of this question. Average total costs, excluding the export tax, remained substantially below prices throughout the 1970s. Variable costs in the short run are assessed by the management at less than 40 percent of total average costs, excluding the export tax. Labor costs are considered a fixed item. The export tax, although varying with output, should not be included in costs because the tax-receiving government is the owner. In 1972, tin prices at 427 cents/lb (1981 dollars) were at their lowest for the decade studied. In that year, average variable costs in Indonesia, taken at 40 percent of total average cost as reported to the ITC, were only 130 cents/lb. Marginal production units within Timah may have had variable costs above price, but they could not have accounted for a significant share of output. Hence, at least during the 1970s, the full-capacity policy cannot have deviated much from a rational pursuit of profit maximization by the government.

Investments and Capacity Expansion

No clear-cut policy has emerged with regard to investment and expansion of Timah's capacity. Overall gross investment between 1969 and 1978 were quite small. At $134 million, investment was two-thirds as large as aftertax profit. On average, gross investment amounted to 7 percent of the overall funds used by Timah and could therefore not have greatly exceeded depreciation. Hence, expansion of production must have mainly resulted from rehabilitating existing, but inactive, installations. In the Indonesian National five-year plan ending in 1984, expansion of Timah's output had been set at 4 percent a year.

Indonesia has a very favorable potential for further growth of its tin industry. Although its share of world output in 1980 was only 14 percent, the country's reserves constituted close to 25 percent of the world total.[35] Given the indus-

34. Interview with management of Timah.

35. P.C.F. Crowson, *Non-fuel Minerals Data Base* (London, Royal Institute of International Affairs, 1980).

try's ability to make sizable fiscal contributions and still generate substantial returns on invested capital, why have the government and its company not pursued a more aggressive investment policy to expand Timah's production even faster? Such a policy would certainly have helped achieve the company's long-run profit maximization objective.

The public ownership of Timah may explain why this corporation grew relatively slowly despite its impressive record of fiscal contributions and profits. The government may have chosen to restrict the growth of the tin and oil sectors, in favor of manufacturing and agriculture, to diversify the economy or better satisfy basic needs, despite the lower returns on investments in the latter activities. The high petroleum income earned through most of the 1970s would have permitted the government to afford such policies.

Another explanation may be the slow growth of tin demand and the geographical concentration of tin production. In addition to the ITC export controls, a tacit understanding may exist among the four major producing countries to keep output in line with demand growth, in a joint effort to maintain favorable tin prices.

Distinguishing Features of Indonesia's Tin Industry

The experiences of Timah and its state-owned predecessors since 1950 provide some support to the hypotheses about state enterprises in international mineral markets.

The account of what happened to Indonesian tin in the first seventeen years of nationhood provides a clear-cut illustration of the hypothesis that nationalization involves heavy start-up and learning costs. The sharp fall in output after nationalization and slowness of the subsequent recovery are primarily explained by the managerial deficiencies and the ensuing inability to expand or even maintain capacity, typical of the inexperienced state enterprise. In two respects, however, the Indonesian experience deviates from the standard story described in chapter 3. First, the chaotic political conditions prior to 1967 and the concomitant financial drain of the industry by the government would appear somewhat exceptional in an international context. The government's appropriation of Timah's revenues in that period undoubtedly contributed to the decline in production and to the difficulty in restoring pre-nationalization production levels. The second unusual feature is the brave and economically not very successful effort to launch a technically complex venture of forward integration into tin smelting, long before the managerial and technical resources had been consolidated.

The activities of Timah in the 1970s also support the hypothesis that state enterprises operate at higher cost levels than would private firms in similar circumstances. The excessive number of employees in Timah and its pursuit of social obligations far beyond those normally required of private enterprise in Indonesia, along with the relatively lax attitude of management and government with regard to cost minimization, provide ample illustrations of this point.

Though costs appear elevated in relation to the minimum attainable, they are quite low in comparison with tin prices. This is mainly a reflection of Indonesia's very favorable tin resource base. In all likelihood, the marginal cost of Timah's operations has not exceeded price for any length of time during the period since 1969, for which cost data are available. The company's policy of maximum feasible utilization of capacity has probably been rational from a profit maximization standpoint. Thus, the Indonesian case study has not been conducive to testing the hypothesis that state enterprises are less flexible than private firms in adjusting capacity utilization to price and market demand.

Nationalization of Indonesia's tin did indeed break up existing chains of vertical integration with Dutch processing units and so, for a time at least, expanded the arms-length sales of tin concentrates. The country's establishment of a smelter in the late 1960s partly reestablished the vertical integration of its tin industry, but not to the extent of the former Dutch operations, which included some tin-related fabricating and manufacturing activities.

Finally, low investment in Timah, along with the reestablishment and growing importance of private foreign-owned tin production in Indonesia since the mid-1970s, conform with the hypothesis that state enterprise is not a survival threat for continued private endeavors in the mineral sector.

Appendix 4.A
Indonesian Interviews

PT Tambang Timah

Ir. M. Simatupang, Director, Research and Development
Mr. Kantakusumah Kusyaman, Head of Marketing Division
Mr. Karim Latief, Head of Training Division
Mr. Asyik Ramly, Head of Mining Division
Iman Mohammed Ismu, Head of Research and Planning Division

PT Riau Tin Mining

Mr. C.B.C. Valk, General Manager

PT Broken Hill Proprietory Indonesia

Mr. C. Mirach, Public Relations Manager

Ministry of Mines and Energy

Dr. Soetaryo Sigit, Secretary General
Dr. J.A. Katili, Director General

Ministry of Finance

Dr. Dono Iskandar, Head of Bureau of Planning

Other

Prof. Mohd Sadli, Former Minister of Mines and Energy
Mr. M. Subroto, Director, General Ministry of Mines and Energy, retired since 1973
Dr. Ralph Beals, Harvard Unit, Advisor to Minister of Finance
Mr. H. Roden, IMF Resident Representative

5

The State Iron Ore Industry in Venezuela

Iron ore has been mined in Venezuela since the early 1950s. The opening of the Venezuelan production units began a wholesale locational shift of the world iron ore industry in the 1950s and 1960s. The depleting iron ore deposits in the United States and Western Europe gave way to the much richer resources of Latin America, Western Africa, and Australia, which had become economically accessible in that period as a result of revolutionary developments in bulk transport techniques. The mines are located in the Guayana region, close to Puerto Ordaz at the confluence of the Orinoco and Caroni rivers. At the time of their establishment, this area was undeveloped and sparsely populated. The foreign investors, in collaboration with the government, had to build up extensive infrastructure for the iron ore operations.

To attract workers to this "frontier" area, the companies had to provide housing, commissaries, schools, and hospitals and to offer salaries substantially above the average for other parts of Venezuela.

The mines were run as two separate enterprises, owned by U.S. Steel and Bethlehem Steel, which wanted to secure their raw material supplies. At first, the Venezuelan units functioned as "captive" mines; the firms consumed most of their own output. The expansion of production in the late 1950s led to an

The data upon which the analysis is based have been obtained from material published in Venezuela and elsewhere, and, above all, from a series of personal interviews, which included a number of office holders in the Ministries of Planning (Cordiplan) and Energy and Mines and in Ferrominera, the state iron ore enterprise, conducted during a two-week visit to Venezuela in November 1982. Appendix 5.A lists the persons interviewed. A particularly valuable source is a detailed memorandum on the state-owned iron ore company, prepared by Janet Kelly Escobar, and entitled "Ferrominera Orinoco," Stencil, December 1979.

increase in arms-length sales. In the 1960s, about a third of the output came to be sold to other steel companies, predominantly in Europe.[1]

The genuine profitability of the iron ore mining activities in Venezuela during the period of foreign ownership is difficult to establish. Until the mid-1960s, the declared profits of the two firms were determined by the intra-company transfer prices that were applied to their internal sales. In the subsequent decade, profitability depended upon the reference prices the government established in an effort to assure Venezuela of its "proper" share of the mineral rent.[2] According to Gomez, the declared aftertax profits varied between 10.2 and 40.2 percent on capital investment in the years between 1957 and 1968, with an unweighted average equal to 22 percent.

In 1975, the iron ore mines were nationalized and have been run since 1976 by a single state-owned enterprise, Corporacion Venezolana de Guayana (CVG) Ferrominera Orinoco CA.

Table 5-1 provides data for Venezuela's iron ore production and exports from 1955, when the industry had acquired significance, and until 1982.[3] Venezuela's share of the world total is also given.

The immediate impulse to nationalize the iron ore industry was founded in the euphoria following the OPEC price increases starting in 1973 and the simultaneous general commodity price boom. The more basic reasons were internal to Venezuela and concerned arrangements in the petroleum industry.

After renegotiations undertaken in the 1940s, the major petroleum concessions in Venezuela were timed to expire in 1983.[4] Uncertainty about the government's intentions caused increasing reluctance among the private petroleum companies, from the early 1970s onwards, to maintain existing installations or to commit funds to create new capacity. Unwilling to accept a gradual dismantling of the country's petroleum capacity, the government decided that the industry had to be nationalized, even before the expiration of the concessions in force.

The petroleum industry is vital to the Venezuelan economy. To minimize the risk of costly mistakes in the nationalization procedure, the government decided the iron ore industry should be nationalized first in order to gain some

1. H. Gomez, "Venezuela's Iron Ore Industry," in R.F. Mikesell et al., *Foreign Investment in the Petroleum and Mineral Industries* (Baltimore, Johns Hopkins University Press, 1971).

2. Ibid.

3. Except where specially indicated, all tonnages in this chapter refer to actual weight.

4. This and the next paragraphs are based on an interview with Argenis Gamboa, retired president of Corporacion Venezolana de Guayana, which is the formal owner of CVG Ferrominera. Gamboa was the person mainly responsible for carrying out the iron ore nationalization.

Table 5-1. Venezuela in the World Iron Ore Industry
(million tons, unless otherwise noted)

	Iron ore production			Iron ore exports		
Year	World	Venezuela	Venezuelan share (%)	World	Venezuela	Venezuelan share (%)
1955	378	8.4	2.2	89.9	7.8	8.7
1960	512	19.5	3.8	151.6	19.3	12.7
1965	624	17.5	2.8	211.7	17.0	8.0
1970	774	21.9	2.8	323	21.1	6.5
1971	781	20.2	2.6	318	19.2	6.0
1972	781	18.3	2.3	311	16.5	5.3
1973	853	22.9	2.7	375	21.9	5.8
1974	899	26.4	2.9	409	26.6	6.5
1975	884	27.0	3.1	389	19.9	5.1
1976	892	18.9	2.1	372	17.3	4.7
1977	843	13.8	1.6	354	11.9	3.4
1978	840	12.6	1.5	348	12.8	3.7
1979	902	14.2	1.6	387	13.0	3.4
1980	877	15.4	1.8	374	11.8	3.2
1981	852	14.9	1.8	360	12.4	3.4
1982	779	10.5	1.4	312	6.6	2.1

Sources: 1955–1975: UNCTAD, "Iron Ore: features of the world market, Statistical Annex," TD/B/IPC/IRON ORE 12/Add.1, August 8, 1977; 1976–1982: APEF, *Iron Ore Statistics*, September 1983.

of the experience needed for an efficient takeover of petroleum. In 1974, it announced its intention to nationalize the two iron ore producers, and the actual takeover was completed by January 1975. The petroleum industry was nationalized in the following year. The officially stated reasons for nationalizing the iron ore industry included appropriating the full mineral rent, placing Venezuelan nationals in the highest managerial positions, increasing national value-added by forward processing of the ore into pellets or briquettes, and assuring the raw material supply to the nascent domestic steel industry.[5]

Though the foreign companies held exploitation concessions valid for many years into the future, they did not contest legally the government's right to nationalize. Negotiations were held to establish the foreign owners' compensation, as well as other terms in connection with the government's takeover.

CVG, a public regional development corporation, was to take over the

5. J. Kelly Escobar, "Ferrominera Orinoco," *Stencil*, December 1979.

companies' assets and be responsible for conducting the negotiations. The agreement finally reached based compensation payments on the net book value of assets, to be paid with interest not exceeding 7 percent a year for the ten years following nationalization.

The negotiations also led to the establishment of several other agreements with the foreign owners:

a. a three-year management and technical assistance contract assuring the continued presence of almost all key foreign personnel for a period after takeover and including a training component to speed up the transition to a fully Venezuelan management,
b. an agreement with U.S. Steel to assist with equipment purchases and with ore marketing in Europe, and
c. long-run delivery contracts to the former principals for virtually unchanged quantities of iron ore for several years following nationalization.

U.S. Steel agreed to take about 11 million tons a year from 1975 to 1981. Bethlehem Steel committed itself to purchase about 3.3 million tons annually from 1975 to 1977. Prices in these contracts were set equal to the price for Mesabi non-Bessemer ore delivered in Pittsburgh or Lower Lake Erie, subject to an agreed minimum.[6]

Contributing to the speedy and smooth completion of the takeover negotiations was the fact that in 1974, the year when they took place, the world steel industry was experiencing a strong boom, of which no end was perceived. The primary concern of the U.S. steel companies at the time was to secure their raw material supply from Venezuela. Their ownership position was a secondary consideration. With their supply needs assured through the long-run delivery contracts, they were more willing to agree to the other terms of the nationalization sought by the Venezuelan government.

The far-reaching collaboration between the old and new owners in the managerial, technical, and marketing fields aroused some political criticism on grounds that the continued foreign presence and involvement diluted national control, one of the key aims of nationalization. The response of CVG was that close collaboration with the former owners was essential to prevent a loss in output and productivity in the period following the ownership transfer.

6. The Mesabi price is a CIF quotation reflecting the special conditions that characterize the U.S. iron ore market. A combination of transfer pricing decisions internal to the U.S. steel companies and to the structure of the U.S. tax system has kept this price at levels substantially above the import prices in Europe (Kelly Escobar, "Ferrominera Orinoco").

The Establishment and Operations of Ferrominera

After a one-year interim period, responsibility for the state-owned iron ore activities was shifted from CVG itself to a fully owned subsidiary, CVG Ferrominera Orinoco CA, established for this purpose in December 1975.

Direction, Control, and Goals

Ferrominera is a fully owned subsidiary of CVG, a public regional development corporation. Other CVG subsidiaries produce steel, aluminum, and hydro-electricity, and engage in other activities. Formally, all responsibilities for Ferrominera are entrusted to CVG, which constitutes the shareholders' assembly and appoints the president and board members. In reality, the lines of overall direction are more complex. For instance, the ministry of energy and mines ensures that Ferrominera adheres to the provisions of Venezuela's mining law. The ministry of planning (Cordiplan) integrates the iron ore activities with the national development plan.

Long-run policies and major investments of Ferrominera are determined in regular meetings between the president and management of the enterprise and representatives of the two ministries. Any disagreements would be settled by the ministry of the secretary of the presidency, the formal principal of CVG. Still other public bodies would be involved if Ferrominera required external capital additions, which it has not so far.

In practice, political involvement in Ferrominera's day-to-day activities seems quite limited. The company's president, along with its top management, appear to have been left in peace in running the company since the time of its establishment. The noninvolvement of politicians could be the result of the manipulating skills of the company's presidents, enabling them to pursue their own ideas by playing off the different political interests against each other. But it could also be because of unanimity between the management and the political institutions about the goals and policies the company should pursue.

A unanimity of opinions on Ferrominera's goals does indeed emerge from my interviews. Both the managers in the enterprise and the responsible ministry officials invariably stated that profit maximization is not a company goal. Although a general consensus exists on the need to avoid losses and to earn some profits, possibly so as to cover the company's own investment requirements, generating surpluses is not a major objective.

A second goal that appears to be mutually accepted is assurance of raw material supply to the domestic steel industry, presumably meaning that deliveries to satisfy domestic needs should have priority over export deliveries. This goal has not had any practical implications so far. In all years since nationalization, the company's capacity has been more than sufficient to satisfy both internal and export market demand without any constraint.

A third goal for Ferrominera, widely professed by the people I interviewed, is that of promoting development in the Guayana region. The foreign-owned iron ore companies were important in regional development, both because they needed infrastructure and social facilities for efficient iron ore extraction and because they wanted to appear as "good corporate citizens." As of the 1970s and 1980s, the Guayana region has been reasonably well endowed in these respects, so continued regional development is less a pressing prerequisite for iron ore production. With the transfer of ownership, regional development ventures projects have gained in importance, even when they do not improve the efficiency of iron ore operations. One of the managers pointed out that since the ultimate purpose of Ferrominera's profits, over and above the company's own financial needs, is precisely regional development, the question of whether the development expenditure is financed by company profits or is considered part of overall costs is not important.

The Record of Production and Sales

Figure 5-1 traces spot quotations for iron ore in the West European market. The price for iron ore is less easy to identify unambiguously than the prices for tin and copper. Most international iron ore trade takes place at prices determined in annual or longer contracts. These prices are incompletely documented, and they can deviate substantially from the spot quotations. Also, large disparities in contract price levels have existed between different geographical areas.

Over longer periods of time, however, spot and contract prices tend to move in parallel. Hence, the figure does reasonably reflect trends in international price levels. It shows that constant-dollar iron ore prices have been falling persistently since the late 1950s. The weakness of prices in the 1960s is primarily explained by the introduction of large-scale and low-cost supply from Latin America, Western Africa, and Australia. The further sharp fall in the latter half of the 1970s reflects the depression in demand caused by the worldwide crisis in the steel industry.

As table 5-1 shows, both production and exports of iron ore in Venezuela reached their peak in the 1970s and have settled at substantially lower levels in

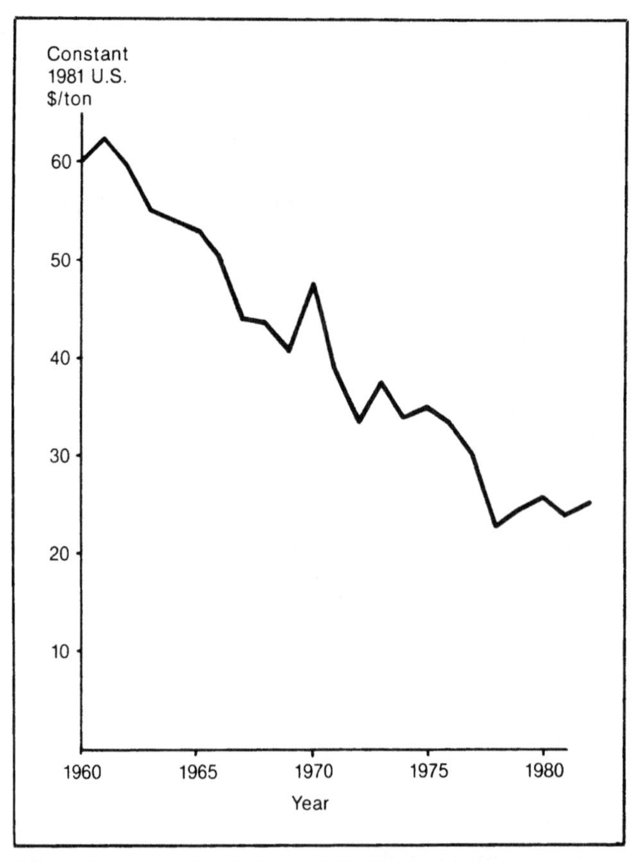

Figure 5-1. Iron Ore Prices, 1960–81 (Brazilian deliveries CIF North Sea Ports, 68% iron content) ***Source*****: World Bank,** ***Commodity Trade and Price Trends*****, 1982/83 edition.**

subsequent years. To an important extent, the decline was the consequence of the exceedingly weak international iron ore market after the mid-1970s. World production and world exports, too, have declined since that time. Closer scrutiny, however, reveals that Venezuela's output and sales fell by much more, with a consequent decline in the country's share in world production and exports.

Table 5-2 records the emergence of a large domestic market for Ferrominera in recent years. Shipments to the national steel industry were insignificant until 1978 because the country's major steel producer, Sidor, was itself operating a smaller iron ore mine (Cerro San Isidoro). In 1980 and 1981, sales in Venezuela reached almost 3 million tons a year, and they rose to nearly 4 million tons in 1982.

Table 5-2. Ferrominera, Iron Ore Shipments
(million tons)

Destination	1974[a]	1975[a]	1976	1977	1978	1979	1980	1981	1982
United States	15.5	11.2	9.8	6.3	6.2	4.8	3.9	5.4	1.2
Europe[b]	10.8	8.2	7.5	5.5	6.6	8.2	7.7	7.0	5.4
Total exports	26.3	19.4	17.3	11.8	12.8	13.0	11.6	12.4	6.6
Venezuela	0.2	0.2	0.1	0.2	0.5	0.7	2.8	2.9	3.9
Total shipments	26.5	19.6	17.4	12.0	13.3	13.7	14.4	15.3	10.5

Note: The difference in the figures of tables 5-1 and 5-2 is partly due to the use of international and national statistical sources, respectively. Production in table 5-1 relates to crude ore output at the mine; shipments in table 5-2 refer to final products sold. The processing to obtain the latter will regularly involve some loss in weight.

Sources: 1974 and 1975, *Hierro y Octros Datos Estidisticos Mineros*, Ano 1980, Ministry of Energy and Mines, Caracas; 1976–1982, figures provided by Ferrominera's accounts department.

[a]Sales for 1974 and 1975 pertain to the predecessors of Ferrominera and are given for comparison.

[b]Includes insignificant amounts exported to Latin America.

The important market loss occurred in the United States, where practically all sales have been based on the long-run contractual agreements with U.S. Steel and Bethlehem Steel. In 1974, these two companies had made commitments to buy about 14 million tons of ore a year between 1975 and 1977 and about 11 million tons from 1978 to 1981. As the world steel depression became apparent, however, the former owners pressured Ferrominera to renegotiate the purchase contracts. Under the new agreements, prices were reduced to the average U.S. East Coast competitive import price and volumes to be purchased were substantially lower. After 1977, the quantities of Venezuelan ore the United States bought were less than half those envisaged in the original contracts. In 1982, the first year after the contracts expired, Venezuelan sales to the United States fell to 1.2 million tons—less than a tenth of the 1974 level.

In 1983, Ferrominera signed a new contract with U.S. Steel, covering annual deliveries of 3 million tons of ore until 1995. Bethlehem Steel has declined the Venezuelan offers of continued supply, pointing to the low utilization of its steel capacity and to commitments to purchase iron ore from Canada and Liberia.

The reduced U.S. interest in the Venezuelan iron ore supply appears to be largely explained by the nationalization. Table 5-3 indicates that imports of ore from Venezuela to the United States declined sharply after nationalization. Imports fell both in absolute amounts and in relation to total imports into the United States. With the rupture of the vertical integration structure, Venezuelan iron ore was assigned a priority below that of domestic and Canadian captive deliveries. As the steel depression developed, ore supplies from Venezuela were among the first to be cut. With continued U.S. ownership of the

Table 5-3. U.S. Iron Ore Market, 1974 and 1978
(million tons, Fe content, unless otherwise stated)

Item	1974	1978
Demand	90.8	80.9
Supply		
Domestic	59.0	60.1
Imported (net)	31.8	20.8
from Venezuela	10.3	4.2
Venezuela's share of		
Demand (%)	11.3	5.2
Net imported supply (%)	32.4	20.2

Source: U.S. Bureau of Mines, *Minerals in the U.S. Economy, annual, several issues.*

Venezuelan mines, the role of Venezuelan ore in the U.S. market might well have been maintained.

The negative impact of the ownership change is confirmed by the comparatively lesser decline in exports to the European market, where supplies from Ferrominera were not similarly down-ranked. Venezuela's share of seaborne imports into Western Europe was 7.3 percent in 1970 and 5.4 percent in 1979.[7] Only in the short run did European sales suffer severely from nationalization, and this was mainly because of a statement by Venezuela's president that in the long run the national enterprise would not export at all. The management of Ferrominera eventually succeeded in convincing the president that continued competition in international markets was essential to assure domestic supply at low cost over time. Initially, however, the uncertainty about continued availability of Venezuelan ore made European customers reluctant to continue their purchases. Considerable persuasion by the company's marketing arm eventually overcame this reluctance.

About half of the Venezuelan ore sales to Europe have been delivered under annually renewable contracts. Most of the remaining quantities have been sold under contracts specifying quantities for five to seven years, with prices renegotiated each year. In practice, these two contractual forms differ little, because the annual contracts are regularly renewed. Some small quantities of ore have been sold in the European spot market. No plans exist to change these marketing arrangements in Europe.

Like its foreign-owned predecessors, Ferrominera has been a price taker in the European market. For many years, the CIF price level in Europe has been

7. J. Husgen, "Sea Transport and Handling of Iron Ore." Paper delivered at Third Bulk Handling and Transport Conference, Amsterdam, May 1981.

determined annually through the outcome of German negotiations, first with Sweden and more recently with Brazil, after that country took over the former Swedish role as leading ore exporter to Europe. Acceptance of this price level has assured Venezuela of a share in the market, even in periods of low demand, because buyers have been interested in diversifying their sources of supply. The buyers' urge to diversify would weaken if some sources of supply were cheaper than others. A unilateral Venezuelan price increase, for example, would entail severe risk of losing sales. In a competitive market, Ferrominera might have considered reducing its prices to assure fuller capacity utilization—a rational profit-maximizing policy given the very low variable costs of operation, fluctuating between $1.8 and $2.4 a ton (17–25 percent of total cost) during the 1977–81 period. With the oligopolistic structure that has characterized the international iron ore market, however, such a policy would have been self-defeating because other producers would undoubtedly have responded by matching price reductions.

Prices for domestic sales have been determined at negotiations between Ferrominera and Sidor, the state-owned steel enterprise. The two firms are sister companies, both owned by CVG, and with several interlocking directorships.[8] Sidor is in close proximity of Ferrominera's port facilities, so the costs of transporting ore from the port are insignificant. Initially, the price levels for deliveries to Sidor were set close to the F.O.B. price Ferrominera received from its European sales. In recent years, Sidor has paid somewhat more than the European F.O.B. price. Given the high transport cost to Europe, the delivered domestic price is substantially below the delivered price to Europe.

Table 5-4. Venezuelan Iron Ore Prices, 1976–82
(U.S. dollars per ton)

Destination		1976	1977	1978	1979	1980	1981	1982
CIF	United States	22.0	20.9	18.9	20.7	24.8	26.9	25.9
	Europe	17.9	17.7	16.0	17.4	22.8	25.0	25.2
FOB	United States	17.8	16.5	14.5	15.5	18.2	19.2	17.5
	Europe	11.4	11.5	9.0	9.5	11.0	11.2	13.1
Sidor		11.7	12.2	11.3	11.1	13.7	14.4	14.4
Weighted average price, excluding ocean shipping		14.8	14.0	11.6	11.7	13.2	14.5	13.3

Source: Ferrominera.

8. Sidor is part-owned by Fondo de Inversiones de Venezuela, FIV, an investment finance agency of the government.

Ferrominera's Capital Structure, Costs, and Results

Ferrominera has been fortunate in starting its life virtually without long-run debt to outsiders. The company's capital structure appears in table 5-5. A substantial part of the long-run debt it does have consists of loans from the Employees' Provident Fund.

Table 5-5. Ferrominera's Capital Structure, 1976–82
(millions of U.S. dollars)

	1976	1978	1980	1982
Short-run debt	68	84	145	133
Long-run debt	32	39	64	72
Own capital	214	218	210	214
Total	314	341	419	418

Source: Ferrominera.

As is evident from table 5-6, Ferrominera has been persistently successful in attaining one of its goals—generating some profits. In 1976, profits before tax provided a return of 4.5 percent on the company's own capital. The return fell to below 3 percent in 1977 and was quite insignificant in the following years. The major reason behind the deteriorating profitability is the worldwide depression in the iron ore industry, which caused low capacity utilization and falling real price levels. Ferrominera's ability to avoid losses since its establishment has been greatly aided by its favorable capital structure and low interest payments.

The sharp variation in "other costs" is primarily due to the losses Ferrominera incurred through its part-ownership in a high-iron briquette plant, in which briquettes containing 85 percent Fe were produced through a natural-gas-based direct-reduction process. This plant, the first of its kind worldwide, has been a perennial loss-maker since it started operating in 1971. The unfavorable results have been primarily due to technical problems that prevented the plant from attaining anywhere near its designed capacity. The installations were closed down in 1977 for redesigning, opened again in 1980, only to be permanently closed in 1982. Ferrominera's net losses on the operations of the briquette plant were $20.1 million in 1976, $3.7 million in 1979, $10.3 million in 1980, and $12.4 million in 1981.

The exceptionally high personnel expenditure in 1976 is explained in the main by the expensive management contracts with the former owners—a temporary start-up cost that declined in the following year. Nevertheless, the

Table 5-6. Ferrominera, Summary Profit and Loss Account, 1976–82
(millions of U.S. dollars)

Item	1976	1977	1978	1979	1980	1981	1982
Gross sales	351.7	232.6	228.7	250.0	311.7	361.7	224.4
Less transport charges	98.1	67.4	86.2	104.8	130.0	153.4	83.9
Net sales	253.6	165.2	142.5	145.2	181.7	208.3	140.5
Costs							
Personnel	77.0	47.4	47.9	54.3	81.3	86.6	62.0
Materials & services	43.0	36.2	42.7	42.2	49.8	55.0	55.1
Depreciation	10.4	10.9	9.7	11.4	13.1	15.0	12.9
Interest paid	0	0.4	4.7	8.7	11.8	15.6	11.8
Other costs, net	26.5	11.2	5.2	7.9	25.0	33.1	(6.8)
Total costs	156.9	106.1	110.2	124.5	181.0	205.3	135.0
Profits before tax	96.7	59.1	32.3	20.7	0.7	3.0	5.5
Tax payments	56.3	34.9	19.6	14.8	0.6	1.7	4.2
Profits after tax	40.4	24.2	12.7	5.9	0.1	1.3	1.3
Dividends	0	0	13.8	14.0	0	0	0

Source: Ferrominera.

level of personnel costs throughout the period studied, at more than 40 percent of total cost, has been very high for a mining operation. To some extent, this is a legacy from the time the mines were established, when labor had to be attracted to Guayana by high wages and a miscellany of company-provided social facilities. Once introduced, these special benefits have continued, both under private-foreign and public-national ownership, despite the infrastructural and social developments in the Guayana region.

After nationalization, CVG tried to pressure Ferrominera to reduce wages and other employee benefits to levels more in line with those offered by its other subsidiaries such as the steel and the aluminum companies. A reason why these pressures have not had much impact so far could be that Ferrominera has always been profitable, while CVG's steel and aluminum ventures have been perennial losers. In 1981, Ferrominera's costs for providing its employees such fringe benefits as housing, hospitals, schools, and a commissary amounted to $17.5 million, 20 percent of overall personnel costs, and almost six times the net profit declared in that year.

Another explanation of Ferrominera's high labor cost has been the sharp fall in labor productivity after the nationalization. The developments depicted in Table 5-7 are less related to the inexperience of the new management that took over after the departure of the foreign owners than to the combination of falling output and social and legal pressures against reductions of the labor force, so long as the company generated profits.

Table 5-7. Labor Productivity in Venezuelan Iron Ore Mining, 1971–82

	1971	1972	1973	1974	1975	1976	1977	1978	1979	1980	1981	Est. 1982
Employees	3,248	3,335	3,372	3,657	3,868	4,012	3,814	3,814	4,222	4,440	3,809	3,300
Tons produced per employee	6,219	5,547	6,850	7,219	6,411	4,710	3,618	3,304	3,363	3,468	3,912	3,182
Labor productivity Index (1971–73=100)	100	89	110	116	103	76	58	53	54	56	63	51

Source: Hierro y otros datos and table 5–1.

Company officials claim that the fall in labor productivity is exaggerated by the responsibility for river dredging operations, which were entrusted to Ferrominera between 1976 and 1981 and added some 200 employees to the company's payroll.

The profit squeeze experienced by Ferrominera since 1980 has led to forceful actions to trim costs. The briquette plant was closed in 1982, reducing the labor force by about 300. In the same year, the dredging operations were transferred out of the company, reducing employment by another 200 and relieving Ferrominera of an excessive proportion of the total costs for keeping the river navigable.

Along with efforts to cut out organizational slack in other areas, the management succeeded in reducing the total labor force from 4,440 at the end of 1980 to 3,300 late in 1982. The cost reductions enabled the company to break even in 1982, but the sharp curtailment in production kept labor productivity low despite the sizable reduction in labor force.

Ferrominera is liable to a royalty equal to 1 percent of the value of production at the mine and, like other companies in Venezuela, to a 60 percent profits tax. The royalty is included in operating costs. Following the legacy established during the time of foreign ownership, the profits for the purpose of profits tax are determined on the basis of a posted price for iron ore, set by the government each year. Since nationalization, the government has followed the recommendations of Ferrominera in setting the posted price, so accounting profits and taxable profits are virtually the same. The profits tax Ferrominera pays also appear in Table 5-6.

The enterprise has no clear-cut dividends policy. Dividends have been declared only twice, but they were pure accounting transactions, aimed at reducing the claims of Ferrominera on CVG, that arose from indemnifications of former owners, initially charged to the iron ore company.

Investment Performance and Production Capacity

Ferrominera has made very few new investments during its existence. Gross investments in the six-year period 1976–81 amounted to $94 million, only marginally above the $83 million depreciated during that time. The investment costs, over and above the amounts depreciated, were financed from retained profits.

Two major investment projects have been completed. The first was the establishment of crushing and screening facilities at the harbor in Puerto Ordaz, aimed at satisfying the product requirements of the European customers. The second was the construction of railway transport facilities between the harbor and the nearby steelworks of Sidor, in response to the increase in domestic shipments.

Ferrominera has made no investments to increase its production capacity, and the maximum production that could be reached and maintained in existing installations is in some dispute. Low levels of output have brought a gradual attrition of equipment at the mines. The average age of the equipment in use is quite high. Output could certainly be raised substantially by increasing the number of hours worked, but major investments in new equipment would be required to attain the peak production of 26 to 27 million tons a year, last reached in 1974 and 1975.

With the wholesale shift in export market orientation, away from the United States and toward Europe, mine production at the earlier peak levels would be pointless without additional processing facilities. Although the former U.S. owners continue to buy the crude mine product and treat it further in their own processing installations, the European customers have increasingly demanded graded material like fines or lumps of detailed specification. The growing importance of the European market, along with the exacting requirements for product quality in that market, motivated Ferrominera's investment in crushing and screening facilities. The theoretical capacity of these facilities is about 20 million tons but may be lower in practice because full capacity utilization has never been tested. The capacity to process, rather than the capacity to mine, may become the effective constraint in Ferrominera when the market for iron ore recovers from its present depression.

A striking feature emerging from my interviews with government officials is the absence of ambition to assure a continuing important role for Ferrominera in the international iron ore market. The managers seem resigned to accepting this view. Certainly, they wish the market would recover and permit fuller use of existing capacity, but they have no plans for expanding, or even modernizing, export capacity. Yet, investments in modernization appear indispensable to maintain Ferrominera's international competitiveness, particularly when the company's ocean shipping facilities are compared with those of CVRD, the state-owned Brazilian iron ore giant.

Venezuelan iron ore is transported by rail to the harbor at Puerto Ordaz, which is 340 kilometers inland along the Orinoco River. There, it is transferred onto ocean-going vessels for transport to North American and European harbors. River conditions permit the entry of ore ships no greater than 60,000 to 80,000 tons. These can be fully loaded only in high waters during the rainy season. In the dry season, the ships must leave Puerto Ordaz half-full in order to get through the shallow passages of the river.

In the 1950s, the typical ships used by the world iron ore industry had a capacity of 10,000 to 15,000 tons. At that time, the scale and efficiency of the newly established bulk transport facilities in Venezuela were well ahead of the rest of the world. Since then, however, the rest of the world has made revolutionary changes in iron ore transport, while the basic design of the Venezuelan installations has remained the same. In 1980, more than a fourth of Brazilian iron ore exports to Europe was transported on ships carrying more than 200,000 tons; three-quarters was shipped on carriers of 100,000 tons or more, loaded and unloaded in tailor-made harbor installations on both sides of the Atlantic.[9]

Ferrominera's nonparticipation in the ongoing technological revolution has gradually weakened its international competitiveness. To avoid uneconomic head-on competition with the Brazilian exporter, Ferrominera has been increasingly concentrating its sales in Europe to the ports in the UK, Belgium, or Italy that have not yet been rebuilt to receive the giant ore carriers. As Brazil continues its aggressive policy of expanding and modernizing its iron ore delivery capacity, and as more ports in Europe and North America are equipped with modern unloading facilities, continued inaction on the part of Ferrominera will inevitably lead to its gradual fade-out from the international market.

Large-scale investments in an ocean harbor and in a railway connecting the harbor with the mines are urgently required if Ferrominera is to play a significant and lasting role in the international iron ore market. To justify such infrastructural expenditure, mining and processing capacity would probably also have to be expanded. Additional investments in marketing would then also be needed to enable the company to sell the increasing volumes.

No such plans are in sight. The management appears to be resigned to a future in which deliveries to the domestic steel industry become Ferrominera's main business and exports assume an increasingly marginal role.

The reasons for Ferrominera's lack of ambition in the international iron ore market are unclear. One frequently expressed argument is that the country's iron ore should be saved for future national needs. Given the size of Venezuela's proved ore reserves and the virtual certainty that the reserve base will increase in response to future exploration efforts, this argument does not sound particularly convincing. Current estimates put reserves at more than

9. J. Husgen, "Sea Transport and Handling of Iron Ore." Paper delivered at Third Bulk Handling and Transport Conference, Amsterdam, May 1981. The U.S. East Coast is lagging behind Western Europe in building harbors that can receive the biggest bulk carriers.

2,100 million tons with an average Fe content of around 60 percent, enough for 80 years' exploitation at present capacity utilization and for 50 years if capacity is increased to 40 million tons a year.[10] A more sophisticated version of the same argument is that exploitation of a wasting resource like iron ore for the benefit of foreigners is not worthwhile at the price levels that have prevailed in the late 1970s and early 1980s. Underlying this proposition may be some notion that the ore in the ground has an intrinsic value that is not fully paid by current prices, or—more pragmatically—that at present prices investment in expanded capacity would not pay. Those who argue along this line apparently disregard the likelihood of rising demand and price, once the steel depression is overcome.

In my opinion, the major explanation behind the unambitious attitudes toward Ferrominera's future international role must be sought at a higher national level. In the 1974–81 period, the petroleum sector has provided the government with huge profits and export earnings. The potential of iron ore has been incomparably smaller. In distinction from Brazil, which has lacked oil export revenues, the Venezuelan government was able to afford to neglect the export potential of iron ore. This, along with the insufficiency of Ferrominera's own earnings to cover the costs of a thorough infrastructural overhaul and expansion, seem to be the main reasons for the company's defensive attitude with regard to its future role in export markets. The differences in behavioral patterns between two state-owned enterprises involved in the same business, explained in the main by the dissimilarity of the respective countries' macroeconomic circumstances, demonstrate the difficulties of generalizing about state enterprise behavior.

Iron Ore Mining Under State Ownership

The establishment of state ownership in Venezuelan iron ore has been coincidental with the start of a prolonged depression in the world iron ore market. The consequences of slack world demand must be distinguished from those arising from state takeover.

Nationalization usually involves start-up and learning costs. The size of these costs depends, among other things, on the economic sophistication of the nationalizing country and on the arrangements under which the takeover takes place. Venezuela is one of the most economically advanced countries in the Third World. In the 1970s, Venezuela's per capita income and educational attainment, measured by enrollment in tertiary training institutions, were about twice as high as in Brazil and Mexico.[11] The foreign iron ore miners had promoted a number of Venezuelans into high-level managerial positions. For these reasons, the qualified manpower availability was less of a constraint than in many other nationalizations of mineral activities in developing countries. Through elaborate management contract arrangements with the former owners, Venezuela took great care at the time of takeover to assure a smooth transfer of the managerial responsibility. The ability to produce iron ore was not significantly affected by the ownership transfer. The temporary additional cost of running the operations with the help of a management contract was noted earlier in the text.

Nationalization broke the vertical integration between Venezuelan iron ore and steel production in the United States and thus ruptured the security of captive markets. The drastic cut in U.S. demand for deliveries from Venezuela was partly a result of the worldwide steel depression, but in large measure it was also because U.S. steel companies placed lower priority on Venezuelan supply. The rupture substantially increased noncaptive sales to Europe as a proportion of the company's total exports. The shipments shift from the United States to Europe involved a loss because prices were generally lower in the highly competitive European market.

This development supports two of the hypotheses of chapter 3. First, Ferrominera did experience start-up costs after nationalization, even though the losses in efficiency because of managerial inexperience were small and of short duration. The disruption of the international marketing arrangements following nationalization required new marketing efforts, which can be seen as part of the start-up costs. The large size of the total start-up cost primarily reflects the peculiar pricing system for iron ore in the U.S. market at that time and the reduction of average prices received as Ferrominera was forced to reduce its sales in that market.

Second, the establishment of Ferrominera and the ensuing changes in export marketing clearly reduced vertical integration in the international iron ore industry and widened the noncaptive international market for this product.

Like many state-owned enterprises, Ferrominera has adopted a mixed set of goals with no clear-cut trade-offs among them. The promotion of social and regional development as corporate goals has involved costs that a private, profit-maximizing firm could have avoided. This supports the hypothesis that the social concerns have raised the cost of producing iron ore. The material presented, however, has not confirmed that the enterprise's pursuit of broader social goals was inefficient.

Although its marginal costs have remained far below the prices received, Ferrominera has sharply cut its capacity utilization in response to the decline in U.S. demand. No attempt has been made to increase the company's European market share through aggressive pricing. A price cut by one seller would have led to matching price cuts by all others, with all worse off than before. A warning example is the attempt by Canadian Quebec Cartier in the late 1970s to double its sales to Germany to 2 million tons by cutting the price. After signing the contract, the German buyers used the agreement to bargain for lower prices from other sellers. Having succeeded in lowering prices, they refused to accept the increased volume of deliveries from the Canadian mine.[12] Ferrominera's behavior supports the working hypothesis that experienced state enterprises are aware of producer interdependence in oligopolistic markets and adjust their marketing policies accordingly, much in the same way as do private enterprises. Of course, the company's unwillingness to increase sales by reducing prices may have been reinforced by the government's conception of iron ore as a valuable resource that should be saved for national needs and not wasted on low-priced exports.

Given its behavior, Ferrominera certainly cannot be accused of having contributed to cyclical price instability. Why it behaved in the way it did remains uncertain. The material presented does not appear to be adequate for a proper test of the hypothesis that state enterprises are less willing than private ones to restrict capacity utilization in the face of falling demand.

Ferrominera's failure to increase production capacity might have alternative explanations. The most straightforward one, equally applicable to public and private enterprises, is that when capacity is underutilized due to slack demand, investments will be directed towards reduction of costs and not towards capacity expansion. In both types of enterprises, investments also tend to decline when profits are low. An alternative explanation is that the lack of investment resulted from the government's desire to integrate Ferrominera vertically with the national economy, while reducing its role as exporter. The investment in transport facilities for domestic deliveries and the absence of investments in expanding or modernizing export capacity fit this explanation. This interpretation, if correct, would be somewhat unusual for state mineral enterprises. It presupposes a national balance of payments strong enough to

12. Information from Ferrominera's sales division, confirmed by Malmexport AB (Stockholm).

permit the government to neglect the export potential of its mineral enterprise. It is too early to judge which of the explanations presented here has the greatest validity.

Whatever the explanation, the company's investment performance in its first seven years of existence has been anything but aggressive, which would tend to support the hypothesis that state enterprise does not threaten the survival of the private mineral industry.

Appendix 5.A
Venezuelan Interviews

Ferrominera:

Garnet Sankarsingh, Manager Administration
Gabriel Medialdea, Manager Operations
Diogenes Chollett, Deputy Manager, Sales
Cesar Alvarez, Superintendent of Mines, Piar Division
Eduardo Boccardo, Public Relations Department

Ministry of Planning (Cordiplan):

Maritza Izaguirre, Minister of State
Carlos Vargas, Director of Planning

Ministry of Energy and Mines:

Concepcion de Moreno, Director of Planning
Carlos J. Salaverria, Advisor, Mining Department
Rafaelo Borghes, Mining Department, Projects Division

Other:

Argenis Gamboa, Private Consultant, Former President of CVG
Janet Kelly Escobar, Professor, Instituto de Estudios Superiores de Administracion
George Kastner, Professor, Instituto de Estudios Superiores de Administracion
Enrique Viloria, Advisor, Petroven de Venezuela

6

The State Copper Industry in Zambia

The geographical area that now constitutes Zambia has been a very important copper producer since the early 1930s, when production started on a large scale. During the past fifty-year period, Zambian copper output has been, on average, about one-tenth of the world total (see table 6-1). Since 1932, the copper industry has completely dominated Zambia's economy. For instance, between 1964 and 1970, this industry's contribution to GDP, government revenue, and export proceeds averaged 44 percent, 59 percent, and 95 percent, respectively. With the lower real copper prices that prevailed between 1971 and 1980, the average contribution of the copper firms fell to 20 percent, 16 percent, and 93 percent, respectively.[1] Even at these lower levels, the dominance of copper and its byproduct cobalt in the national economy is overwhelming. In an international perspective, Zambia is extremely dependent on one industry and one product.

The early developments of Zambian copper have been described in detail elsewhere.[2] At the time of Zambia's independence in 1964, British, South

The data upon which the following analysis is based have been obtained from material published in Zambia and elsewhere, as quoted, and above all from a series of personal interviews during a two-week visit to Zambia in January 1983. Among those interviewed were representatives of the National Commission for Development Planning; the Ministry of Mines; ZIMCO, the state industrial holding company; and a number of members of the management team in the Zambia Consolidated Copper Mines (ZCCM), the 60 percent state-owned sole copper-producing company in the country. Appendix 6.B lists the people interviewed.

1. *Zambian Mining Yearbook*, 1973, 1980.

2. See, for instance, S. Cunningham, *The Copper Industry in Zambia: Foreign Mining Companies in a Developing Country* (New York, Praeger, 1981).

Table 6-1. Zambia's Share in the World Copper Industry

Year	Mine production of copper: World (thousand tons)	Mine production of copper: Zambia (thousand tons)	Zambia's share of world production (percent)
1935	1,467	146	10.0
1940	1,992	265	13.3
1945	2,170	199	9.2
1950	2,530	298	11.8
1955	3,130	359	11.5
1960	4,242	576	13.6
1965	5,070	696	13.7
1969	5,935	755	12.7
1970	6,335	677	10.7
1971	6,477	636	9.8
1972	7,063	701	9.9
1973	7,513	683	9.1
1974	7,670	710	9.3
1975	7,317	648	8.9
1976	7,843	712	9.1
1977	7,965	659	8.3
1978	7,846	654	8.3
1979	7,925	584	7.4
1980	7,862	611	7.8
1981	8,310	568	6.8
1982	8,215	530	6.5

Sources: Zambian production 1969–81 from ZCCM; other figures from *Metal Statistics*, Annual.

African, and U.S. interests controlled the country's copper industry, and the Anglo American Corporation Group and the Rhodesian Selection Trust Group managed it.[3]

About the time of independence, copper prices rose sharply and remained high for almost ten years. In 1965, Rhodesia announced its Unilateral Declaration of Independence, which led to a closure of the border with Zambia. For several years thereafter, the copper industry experienced very severe transport difficulties, until the Tazara railway was completed at the end of the decade. To assure uninterrupted operations, the copper companies astronomically increased their inventories, and for a period copper had to be air-lifted to Dar-es-Salaam. Despite the ensuing higher costs, the years after independence were very profitable for Zambia's copper industry, and it contributed substantially to government revenue.

3. A.D. McMahon, *Copper, A Materials Survey* (Washington, U.S. Bureau of Mines, 1965).

Beginning in 1969, the government took several steps that gradually extended its ownership and control over the copper industry. Two major motivations lay behind the desire to nationalize:

1. Public ownership of the means of production was in line with the socialist political philosophy generally pursued by the government.
2. The copper sector was so important to the national economy that the Zambians considered foreign ownership and control politically intolerable. National control of the copper industry was seen as a necessary prerequisite for genuine economic and political independence.

Subsidiary motivations for expanding state involvement in copper included the desire to increase the Zambian share of the mineral rent the copper industry generated, to expand the opportunities for Zambians in the management of the copper companies, and to make the industry more responsive to national needs.

The following have been the main steps in the nationalization process:[4]

In 1970, the government acquired 51 percent of the equity in the copper industry. The takeover involved a reorganization of the industry into two corporations: Nchanga Consolidated Copper Mines, Ltd. (NCCM) and Roan Consolidated Copper Mines, Ltd. (RCM). The government equity was ultimately held through the Zambia Industrial and Mining Corporation (ZIMCO), a 100 percent government-owned holding company entrusted with the responsibility of coordinating the government ownership interests in the industry. Payment for the acquisition was in the form of interest-bearing bonds. The former owners, Anglo American Corporation and Roan Selection Trust, Ltd. (RST), remained minority shareholders and were entrusted with the managerial responsibility for continued operations through elaborate management contract arrangements.

In 1974, after a premature redemption of the bonds issued in payment for the initial equity takeover, the Zambian government cancelled the management contracts with Anglo American and RST. As a result, NCCM and RCM became self-managing companies, with the chief executives appointed by the government. About the same time, the government also established the Metal Marketing Corporation of Zambia, Ltd. (MEMACO), 100 percent owned by ZIMCO and entrusted with the responsibility of marketing the entire output of NCCM and RCM.

4. Government of Zambia, *Third National Development Plan 1979-1983*, Lusaka, October 1979; *Zambian Mining Yearbook 1980*; and ZCCM *Annual Report 1982*.

In 1979, the government increased its equity share in NCCM and RCM from 51 percent to fractionally more than 60 percent. This came about as a result of converting into equity a number of government loans extended to the companies to permit continued operations in the difficult period of low copper prices since 1975. In compliance with the companies' articles of association, the private foreign equity holders were given the offer, but declined, to contribute additional capital to keep their equity share unchanged.[5]

In 1982, finally, the government merged the two companies into a single corporation, Zambia Consolidated Copper Mines, Ltd. (ZCCM), in which it retained a 60 percent equity holding. The last step did not involve any increase in government ownership or control of the copper industry.

Zambia's Copper Industry Since 1970

The direction and goals of the Zambian copper industry in the postnationalization period have been shaped to a considerable degree by three intertwined circumstances. First, the industry is of overwhelming importance to the national economy. Second, in 1964, when the country gained its independence, qualified Zambian manpower was virtually non-existent. At that time, only eighty-nine nationals had university degrees; of these, only three worked in the copper industry.[6] Even in 1970, when the government took over the majority of the equity, the dearth of educated and experienced Zambians was so pronounced that a reasonable functioning of the industry was totally dependent on large-scale expatriate managerial and technical inputs. Third, unlike the Indonesian and Venezuelan cases, nationalization was not total and, because the continued presence of the minority owners was seen as essential, the direction and goals of the industry had to give due consideration to their interests.

Thus, the decisions in 1970 to nationalize only partially, to seek an amicable agreement with the foreign owners who were asked to cede part of their rights, and to request them to continue their management responsibility must be seen in the context of the importance of the industry to the national economy. The experienced foreigners had to stay in order to avoid the risk of wrecking the boat. The first step in the nationalization process had a limited impact on the fundamental policies of the industry. For this reason, it may be appropriate to study the direction and goals of the copper industry as they have

5. Interview with J. Mapoma, Director General of ZIMCO.

6. Interview with D.A.R. Phiri, former Managing Director of RCM.

evolved after 1974, when the management contracts were discontinued and the government appointed two Zambian managing directors. Even in the post-1974 period, however, the circumstances spelled out in the preceding paragraph continue to exert a heavy influence on Zambian copper industry policy formulation.

Since their establishment, the copper companies have formally been subsidiaries to ZIMCO, the state-owned industrial holding company. In reality, the government has usually directly exercised its ownership rights. The three government bodies most immediately involved have been the Ministry of Mines, the Ministry of Finance, and, to some extent, the National Commission for Development Planning. The first has probably played the most important role. Thus, during a number of years, the Minister of Mines and his permanent secretary were chairman and member, respectively, on the two companies' boards. Even though at times the three government bodies have not been in full accord—for instance, in production expansion plans or tax versus dividend policy—the overall coordination has been assured by implicit participation of the Zambian president in formulating key policies or in board and top-level managerial appointments for the sector. Involvement in the industry's affairs by the highest political authority is not surprising, given the importance of copper in the national economy.

In practice, the size of the enterprises, the presidential support their chief executives enjoyed, and the technical competence of their staff have given the copper industry considerable independence and freedom of action, not only in day-to-day affairs but also in long-term policy decisions.

The goals of the nationalized copper sector have never been fully spelled out. Judging from discussions with representatives of ZCCM and the government, profit maximization may be a key goal, but due consideration is also given to employment. None of my interviewees, however, was able to explain the operational meaning of the employment objective. One reasonable interpretation, built up from scattered statements made by several of the managers, is that management has discouraged the introduction of labor-saving innovations and the government will not accept closures of installations that would result in large-scale labor redundancies, unless alternative employment opportunities for the redundant labor could be provided. The greater importance afforded to profit maximization, as compared to PT Timah and Ferrominera, can be seen as a direct consequence of the desire to retain foreign ownership in the industry. The foreign equity holders would probably not stay on, if the profit goal were subordinated to other objectives.

Other objectives receiving attention, but not clearly spelled out in operational terms, include Zambianization of managerial and skilled personnel cadres and maximization of the industry's (presumably net) foreign exchange earnings.

The loose definition of the goals other than profit maximization, and the absence of well-defined tradeoffs among them, leaves the management of the industry with very wide discretionary powers.

Production and Exports

Even though copper dominates the output, it is not the sole product of the Zambian copper industry. ZCCM operates a lead and zinc mine (formerly owned by NCCM), and several byproduct metals, including cobalt, selenium, gold, and silver, are extracted in the copper operations. Cobalt is by far the most important among the byproducts. Zambia's production of this metal in recent years has accounted for approximately 10 percent of the world total. Table 6-2 gives a summary of the volume and value of the different products in recent years. 1979 and 1980 have not been included because the exceptionally high cobalt prices in these two years give a distorted picture of this metal's importance in total sales. Copper normally accounts for 90 percent or more of overall production value.

Domestic consumption of refined copper in Zambia is very small. It reached a maximum of 4,000 tons in 1974 and hovered between 2,000 and 3,000 tons a year in the late 1970s. This corresponds to 0.5 percent or less of output.[7] Except for inventory variations, therefore, production and exports do not differ significantly in the Zambian copper industry.

Production of copper in Zambia rose from 300,000 tons in 1950 to a peak of 755,000 tons in 1969, when the decision to nationalize was taken. This was in parallel with world output. At the time of takeover, the prospects for continued output expansion were viewed quite optimistically. A long-term projection published in Zambia's First National Development Plan (1966-70) arrived at a production level in 1976 of 1.2 million tons. The Second National Development Plan, published in 1972, set the 1976 production objective at 900,000 tons, 30 percent higher than actual output in 1972. This more detailed forecast was arrived at after consultations with the foreign managements of the two copper companies, and it took the planned investments in the copper sector into account.

In actual fact, the 1969 production peak was never again attained. The absolute output levels since that year show a falling trend. The result has been

7. *Metal Statistics 1982* (Frankfort am Main, Metallgesellschaft).

Table 6-2. The Zambian Copper Industry: Output and Sales Value of Metals

	Annual Average 1974–78		Year ending March 1981		Year ending March 1982	
Metal	Output (000 tons)	Share of value[a] (percent)	Output (000 tons)	Share of value[a] (percent)	Output (000 tons)	Share of value[a] (percent)
Copper	674	92	598	89	570	89
Cobalt	1.7	3	1.3	5	2.3	6
Lead and zinc	62	4	44	2	45	3
Other metals	—	1	—	4	—	2
Total	—	100	—	100	—	100

Sources: Republic of Zambia *Monthly Digest of Statistics*, January/March 1982; ZCCM, Annual Report 1982.

[a] Based on current selling prices.

a significant decline in Zambia's share of world output.

Was there a causal relationship between nationalization and the sharp turnaround in the output growth trend, especially since the prospects for future production expansion looked so favorable? In all likelihood such a relationship does exist. A closer scrutiny, however, reveals several factors, completely unrelated to the national takeover, that contributed to the sharp deterioration in output performance.

The first factor was the serious accident at the Mufulira mine in September 1970, which reduced production in the following year by almost 100,000 tons. Prior to the accident, the plan was to increase this mine's capacity from 165,000 tons a year, attained in 1969, to 190,000 tons by 1973. The accident resulted in a change of mining methods, and the former capacity expansion plans had to be revised on technical grounds. A result of this change was that by 1976-77, when rehabilitation had been completed, output did not exceed 142,000 tons.

The second factor restraining output expansion was a gradual realization, during the 1970s, that the geology of the Zambian copperbelt was not conducive to economic copper exploitation at much above the 700,000 ton level.[8] Expansion beyond this level would result in sharply increasing investment costs per ton of additional production capacity. The intricately shaped underground ore bodies could not be exploited much above current rates without large-scale new investments to create additional access routes. Also, faster

8. Interview with C. Belshaw, Consulting Mining Engineer, and P. Freeman, Consulting Geologist, with ZCCM.

depletion of the discrete ore formations would result in an uneconomically high rate of depreciation. According to my interviewees, the large capacity expansions projected in the first and second national development plans must have subsumed the discovery of new ore bodies and the development of new mines. For instance, the high production forecasts of both development plans included the anticipated output from the Lumwana deposit off the copper belt, containing upward of 1 billion tons of ore with about 0.9 percent copper content. In fact, no large-scale economic copper deposits were identified in the 1970s, and no new copper mines have been opened. On closer scrutiny, the Lumwana deposit proved uneconomical, especially because of its very heavy infrastructural investment requirements. Retrospectively, therefore, the projections made in the development plans appear grossly overoptimistic.

The third factor has had to do with the weakness in the copper market and the low price levels that prevailed after 1974. Figure 6-1 depicts the development of copper prices in real terms since Zambia's independence. The average price level after 1974 has been less than half the average attained in the preceding ten years. The predominant contractual form under which copper from Zambia is sold is a one-year contract with prices closely related to those of the London Metal Exchange. Even though the demand for copper from Zambia has never been a binding constraint (additional quantities of the high-quality Zambian output could always have been sold at the going price on the LME), the willingness and ability to expand supply was severely dampened after 1974 by low price levels, which reduced profitability and the cash flow required for investments. Thus, actual capital expenditure in the 1972–76 period turned out to be almost 20 percent below plan.[9]

Sharp reductions in the investment plans in the copper industry were not isolated to Zambia but occurred worldwide. In 1974, the copper mining capacity of the entire Western World was projected to increase from an actual figure of 6.7 million tons in 1973 to 9.1 million in 1978, or by 6.2 percent a year. The factual 1978 achievement was 7.5 million tons, implying an annual growth of no more than 2.5 percent.[10]

The factors discussed above obviously explain part of the discrepancy between production plans and actual achievements in the Zambian copper industry, as well as the country's falling share in Western World output, but other factors, related directly to nationalization and to increasing bureaucratic con-

9. Government of Zambia, *Third National Development Plan.*

10. M. Radetzki and C. Van Duyne, "The Response of Mining Investment to a Decline in Economic Growth: The Case of Copper in the 1970's," *Journal of Development Economics*, 1984.

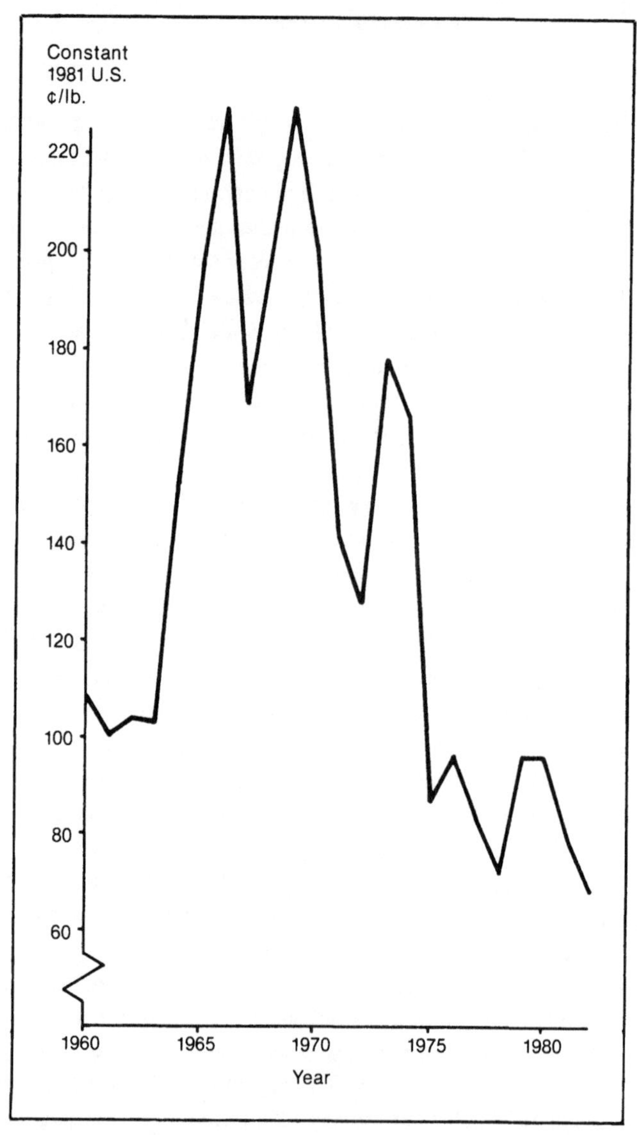

Figure 6-1. Annual Average Copper Prices, 1960–82 (LME Wirebars cash) ***Source*****: World Bank,** ***Commodity Trade and Price Trends*****, 1982/83 edition.**

straints in independent Zambia, also contributed to these detrimental developments. The relative weights of these factors are difficult to assess.

Copper Industry Capital Structure, Profitability, and Cost Developments

Most of the data presented below have been worked out, in response to my requests, by the accounts department of ZCCM. The 1981 and 1982 figures are based on the ZCCM annual report for 1982 and refer to the accounting years ending March 31. A majority of the figures for earlier years have been obtained by adding together the accounts of NCCM and RCM, whose accounting years end on March 31 and June 30, respectively. For simplicity, I refer in the following to the accounting years as ending in March, the terminal accounting month for the bigger of the two companies. The figures, expressed in Zambian kwacha, have been converted to U.S. dollars at the official exchange rates that prevailed on March 31 if accounting years are used, and on December 31 if calendar years are used. The World Bank's U.S. dollar Manufacturing Unit Value Index CIF has been used as a deflator to obtain constant 1981 dollars.[11]

The first set of data, displayed in table 6-3, provides information on the Zambian copper industry's capital structure. The state-owned enterprises started their existence in a very solid position, with no short-term debt and with long-term indebtedness representing less than 10 percent of total funds employed. The industry's solidity remained high through the high-price period for copper, but beginning in 1975, when prices fell, the companies started to borrow heavily to compensate for their sharply reduced cash flow. Although before 1975 practically all their loans were with credit institutions in Zambia, from then onwards they have borrowed substantial amounts abroad. Foreign borrowing amounted to $375 million in 1976 and to $412 million in 1982.

Table 6-4 presents a summary profit and loss account for the copper industry. It has not been possible to subdivide the aggregated "cost of sales" item into all its major components. Furthermore, over time, "depreciation" lacks consistency. After 1978, depreciation has been charged against the operating results in accordance with standard procedures in the international mining industry. Until 1978, however, expenditure on replacement of fixed assets was included in its entirety in cost of sales for the year. The impact of inflation should be taken into account; the values of the early 1970s must be divided by a factor of 2 to 2.5 to make them comparable with the values of the early 1980s.

11. See chapter 1.

Table 6-3. The Zambian Copper Industry's Capital Structure, 1970–82
(millions of dollars)

Item	Accounting Years Ending March 31												
	1970	1971	1972	1973	1974	1975	1976	1977	1978	1979	1980	1981	1982
Short-term debt	0	0	0	0	0	99	291	286	182	212	167	166	179
Long-term debt	46	50	111	153	183	285	360	314	244	267	234	337	502
Own capital	585	707	771	942	1,034	1,113	1,099	911	870	1,209	1,314	1,349	1,033
Total funds employed	631	757	882	1,095	1,217	1,497	1,750	1,511	1,296	1,688	1,715	1,852	1,714

Source: Zambian Consolidated Copper Mines, Ltd. (ZCCM).

Table 6-4. Summary Profit and Loss Account of the Zambian Copper Industry, 1971–82
(millions of dollars)

Item	Accounting years ending March 31											
	1971	1972	1973	1974	1975	1976	1977	1978	1979	1980	1981	1982
Gross sales	935	755	936	1,493	1,161	859	1,028	807	1,116	1,329	1,312	1,061
Cost of sales	531	531	651	718	884	949	945	882	990	1,045	1,270	1,247
Depreciation	—	—	—	—	—	—	—	45	63	63	77	89
Interest payment, net	1	3	9	11	12	39	42	40	48	39	50	72
Devaluation loss	—	—	—	—	—	—	68	33	—	—	—	—
Profit before tax	404	224	285	775	277	−90	83	−75	126	284	42	−186
Tax paid (or recovered)	200	60	81	477	136	−84	54	−48	13	112	−26	3
Profit after tax	204	164	204	452	141	−6	29	−26	113	172	68	−189
Dividend	102	80	105	172	37	0	0	0	6	29	11	0

Source: Zambian Consolidated Copper Mines, Ltd.

The copper industry's profitability has been closely correlated with copper prices. Until 1975 (March), pretax profits yielded a return on shareholders' funds of between 25 percent and 75 percent. Since then, profits have been depressed or negative most of the time. The exception is 1980, when exploding cobalt prices enabled the companies to attain a pretax yield on shareholders' funds at 22 percent.

Tax payments during the twelve-year period covered by the table amounted to a total of $978 million. This constituted 46 percent of net pretax profits, considerably below the nominal 73.05 percent derived by a combination of Zambia's mineral tax and corporate income tax, often quoted in the literature.[12] Restitution of taxes during years of losses has resulted in total net tax payments in the 1976–82 period of no more than $24 million (equal to 13 percent of net pretax profits during that period). Aftertax profits, net, for the twelve years covered by table 6-4 amounted to $1,326 million, of which $542 million (41 percent) was distributed as dividends to equity holders. The share of dividends in aftertax profits declined from 43 percent in the high-profit period, 1971–75, to 29 percent in the low-profit period, 1976–82.

Given its aggregate nature, the summary profit and loss account presented in table 6-4 does not permit much analysis of development of costs. The companies have long separated out the costs attributable to copper, however, and subdivided them by stage of production. Presumably, these costs do not include company administration at headquarters, central services functions, sales commissions to MEMACO, and interest charges.[13] The details of the data given, converted into current U.S. cents/lb, are presented in table 6-5, along with summary series in constant (1981) U.S. cents/lb.

In the judgment of ZCCM's accounting division, about two-thirds of the free-on-rail costs are fixed in a one-year time perspective. It follows that variable costs, constituting one-third of free-on-rail costs plus the cost of transport, have been substantially below the LME prices during the period covered in table 6-5. On their own, the companies have not had any economic reason to reduce production below the technical capacity limit. At midyear 1982, that is, in the 1983 accounting year, which is not shown in the table, copper prices deteriorated further and stayed at a historically low level for about half a year. Though the variable costs for copper production in ZCCM as a whole remained below average revenue in that period, the price proved

12. See, for instance, S. Cunningham, *The Copper Industry in Zambia: Foreign Mining Companies in a Developing Economy* (New York, Praeger, 1981) p. 192.

13. World Bank, "Zambia—A Basic Economic Report, Annex I: Mining Sector Review," October 14, 1977.

Table 6-5. The Cost of Producing Zambian Copper, 1971–72

Item	Accounting years ending in March 31											
	1971	1972	1973	1974	1975	1976	1977	1978	1979	1980	1981	1982
Copper output (000 tons)	745	644	719	690	697	666	689	648	624	537	588	592
Cost (current U.S. cents/lb)												
Mining	14.4	15.6	16.7	19.4	24.1	25.8	23.5	25.1	26.7	31.7	37.6	34.6
Concentrating	2.7	3.0	3.6	3.9	5.0	5.6	4.3	5.2	5.8	8.6	9.2	13.2
Smelting, leaching & refining	4.6	5.4	6.5	7.0	8.9	9.9	9.2	8.5	11.7	13.9	16.3	18.6
Administration and general	4.5	4.7	5.0	6.7	7.5	7.7	7.2	8.8	9.3	11.3	12.6	11.2
Total cost, free on rail	26.2	28.7	31.8	37.0	45.5	49.0	44.2	47.6	53.5	65.5	75.7	77.6
Transport costs (preceding calendar year)	4.0	4.1	4.8	4.8	5.5	6.1	4.9	4.7	5.6	6.0	7.5	n.a.
Total costs CIF	30.2	32.8	36.6	41.8	51.0	55.1	49.1	52.3	59.1	71.5	83.2	—
Cost (constant 1981 U.S. cents/lb)												
Total cost, free on rail	72.8	73.6	67.7	62.7	68.9	72.0	60.5	54.7	55.2	62.4	75.7	77.2
Transport costs (preceding calendar year)	11.1	10.5	10.2	8.1	8.3	9.0	6.7	5.4	5.7	5.7	7.5	—
Total costs CIF	83.9	84.1	77.9	70.8	77.2	81.0	67.2	60.1	60.9	68.1	83.2	—

Source: Zambian Consolidated Copper Mines, Ltd.; Zambia Mining Yearbook.

insufficient to cover the variable costs of some of the producing units in the company. Production was continued at full technical capacity, however, partly because of the social concern for employment. In addition, however, the economics of output reduction were uncertain, given the expense of closing and reopening the high-cost units and the expected short duration of the extreme price trough.

Even though the incompleteness of the data makes it difficult to draw any far-reaching conclusions on the development of costs, the available figures, in combination with the interview results, permit several pertinent observations. The low cost levels in 1977 through 1980 are explained primarily by a devaluation of the Zambian currency in mid-1976. The items included in the total free-on-rail cost, when measured in constant-dollar terms, increased by less than 5 percent from 1971–72 to 1981–82. At first glance, this appears to be an impressive achievement, given the continuous decline in the ore grades mined and the increasing depths from which the ores were extracted, even though the devaluation was of help in containing the dollar-denominated cost increase.

A point many interviewees stressed heavily, which may have contributed to the apparently favorable cost development, is the gradual neglect during the period of low copper prices of maintenance and development expenditures in mines and plants. To limit the fall in declared profits or to reduce emergent losses, the companies have tended to cut those costs that, although indispensable in the longer term, could be deferred without impairing current production. Added to the problem of keeping up maintenance and development expenditures have been the restrictions of the Central Bank in allocating needed foreign exchange to the copper firms. Since 1975, therefore, the industry has operated with an increasing burden of potential expenses to make up for the accumulated negligence of earlier years. Even though it has not been possible to quantify the amounts involved, a significant jump in cost levels is inevitable when that negligence is rectified.

The bureaucratic control of foreign exchange is also asserted to have led to an immediate reduction in efficiency. Allocations of foreign currency have been both restricted and delayed, thereby causing such operational disruptions as the unavailability of spare parts. On many occasions, the Central Bank directed the companies' imports to specific countries by providing only a particular type of currency. This use of exchange controls to discourage imports from South Africa has forced the companies to redirect their purchases to higher cost sources in other more distant countries. Apart from the direct increase in purchase costs, the Central Bank policy has delayed delivery times and required a greater diversity of equipment, thus increasing the cost of spare parts inventories; a wider assortment of parts is needed to keep the equipment operational and, with more distant sources of supply, larger stocks of each item are required.

Both independence and nationalization have involved substantial changes in the labor situation in the copper industry. One key issue has been Zambianization. Another has been that of simultaneous substantial declines both in labor productivity and in real wages.

Table 6-6. Labor Force in the Zambian Copper Mines

Year	Zambian locals	Expatriates	Total
1965	39,680	7,040	46,720
1970	44,090	4,380[a]	48,470
1975	52,990	4,500	57,490
1980	55,260	2,490	57,750
1982	54,230	2,050	56,280

Source: Zambia Mining Yearbooks, several issues.

[a] This figure is exceptional and thus not representative of the trend. Both in 1969 and 1971 the number of expatriates was about 4,750.

Table 6-6 shows the extent of Zambianization of the copper labor force. The Zambian government and the national management teams that took charge of the two copper companies in 1974 have had an ambivalent attitude toward the expatriates working in the industry. On the one hand, the skills and experience of the foreigners are admittedly indispensable to assure efficient operation. On the other, the Zambians have always wanted to reduce their dependence on foreign manpower and to Zambianize at the fastest possible rate. At times, they have tried to slow down foreign exodus. At other times, national and corporate policies have consciously aimed at speeding up the process. On some occasions, both aims have been pursued simultaneously.

Expatriates have been departing in large numbers ever since independence. Initially, the exodus was primarily motivated by the political uncertainty the foreigners felt about their own long-term role in independent Zambia. Later, the main reason became the gradual deterioration in their contract and living conditions. Under the circumstances, recruitment of new professionals to replace those leaving became difficult, and the experience and quality of the professionals that could be attracted gradually declined. The desire to Zambianize explains the absence of clear-cut efforts to reassure the foreigners and stop the deterioration in their remuneration.

By the early 1970s, the lack of experienced expatriates started to affect production in an important way. The problem has been most severe among supervisors and skilled craftsmen. The versatile departing expatriates were replaced by young Zambians with little practical experience, just out from crash-training programs that equipped them with a very narrow range of skills. To manage the work routines, supervisory positions were doubled or trebled in many cases. Additional supervisory layers were also established. Another consequence of the change was deterioration in the standards of equipment maintenance, resulting in increasing cost levels to keep the equipment working. These difficulties are said to have continued into the early 1980s. Both Zambian and expatriate interviewees contend that twenty or more years may be needed to bring up a generation of skilled and efficient Zambian craftsmen, supervisors, and technicians.

The problems of Zambianization have not been as serious at higher technical levels, mainly because relatively few positions in these categories have been Zambianized. In contrast, Zambianization of high-level commercial, administrative, and personnel functions has been quite common and does not appear to have had detrimental consequences for operating efficiency.

The top management of ZCCM would like to stabilize the number of expatriates working with the company at about the current level. To limit further exodus, the contracts offered to expatriates were considerably improved in 1982, thus increasing the relative international attractiveness of employment opportunities in Zambia, for the personnel categories needed.

The reduction in labor productivity resulting from a combination of increased labor force and stagnant or falling production is shown in table 6-7. Interestingly, the large expansion in the number of employees occurred in 1973 and 1974, when high copper prices undoubtedly reduced the pressures to contain costs. After 1974, changes in employment levels remained within a narrow range.

Productivity fell from 14.2 tons of copper per man year in 1969–71 to 9.9 tons in 1980–82, or by 30 percent. Three factors explain this decline. First was the lowering in ore grades, which required the treatment of larger volumes of material per ton of metal. In 1969, 43 tons of ore were treated per ton of finished copper; by 1981, the figure had risen to 52 tons.[14] The effect of this change on labor productivity was probably countered to a substantial extent by technical improvements in the equipment brought into use during the period and by increasing mechanization. The second factor affecting productivity was the impact of Zambianization. Third was the strong social and political pressure to avoid labor force reductions. One of the chief executives of ZCCM stated that precisely because of the importance afforded to employment he was

14. Zambia mining yearbooks and ZCCM *Annual Report*, 1982.

Table 6-7. Labor Productivity and Labor Costs in the Zambian Copper Industry, 1969–82

Item	Calendar years 1969	1970	1971	1972	1973	1974	1975	1976	1977	1978	1979	1980	1981	1982
Total employment (000)	48.2	48.5	49.8	50.9	52.8	56.1	57.5	57.1	59.1	56.7	55.5	57.7	58.2	56.3
Labor productivity (tons of copper per employee)	15.7	14.0	12.8	.	12.9	12.7	11.3	12.5	11.2	11.5	10.5	10.6	9.8	10.9
Total employee earnings														
Current dollars (million)	127	132	143	157	190	217	220	222	229	221	241	278		
Constant 1981 dollars (million)	425	399	399	399	405	369	329	326	310	254	249	265		
Constant 1981 dollar index (1972 = 100)	107	100	100	100	102	93	83	82	78	64	62	66		
Average employee earnings														
All employees, current dollars				3,120	3,690	3,720	3,600	5,040	4,640	3,640	3,930			
Zambian employees, current dollars				2,240	2,630	2,640	2,290	3,510	3,270	2,880	3,320			
All employees, constant dollar index (1972 = 100)				100	99	80	68	93	79	53	51			
Zambian employees, constant dollar index (1972 = 100)				100	97	78	60	90	78	58	60			

Sources: Zambia Mining Yearbook, several issues; *Zambia Monthly Digest of Statistics*, January 1977 and January–March 1982; ZCCM.

not seriously concerned about falling productivity in the copper industry since national takeover.

Although the labor force increased during the period considered, the direct cost of labor was substantially reduced, both in absolute amounts, when expressed in real values, and as a share of total cost. Measured in constant dollars, total employee earnings for 1978–80 were 37 percent lower than for 1969–71. The direct labor costs were about 27 percent of the total cost of sales in the 1972 accounting year; in the 1981 accounting year, the figure is only about 22 percent.

Table 6-7 also traces the development of average employee earnings between 1972 and 1979. The figures show that the fall in labor cost is explained only to a minor degree by the substitution of Zambians for expatriates. The constant-dollar unit cost to the companies for hiring both Zambian and expatriate labor fell by 49 percent during the seven-year span covered by this series of data and by 40 percent if only Zambians are considered.

Expressing employee earnings, as well as other cost figures for the copper industry, in dollars appears to be the relevant approach for an industry in which all earnings and most costs are incurred in convertible foreign currencies. What the figures say is that in one important respect the international cost competitiveness of Zambia's copper industry improved substantially.

What counts from the employees' point of view, however, is real earnings in local currency, and these too declined. For instance, average earnings of all employees in the industry expressed in constant kwacha, fell by 35 percent between 1972 and 1979. The reduction in average real wages in Zambia was not an isolated phenomenon for the copper industry, even though it was more dramatic than in others. The average earnings of all employees in all industries fell by 21 percent between 1972 and 1979, when measured in constant kwacha. The substantial reductions in average real earnings of all employees in Zambia reflect the dominance of copper in the national economy. Mainly as a result of the lower copper prices since the mid-1970s, the country's real per capita GDP declined by 21 percent between 1972 and 1979, which was the major reason for the general downward adjustment in wages and salaries.

Investment Performance

Gross capital expenditures in the Zambian copper industry are shown in table 6-8. One way of judging whether they are high or low is by relating them to the industry's total funds. This is done in column 3 of the table for the period of state ownership. Although the investment series stretches back to 1961, the

Table 6-8. Investment in the Zambian Copper Industry, 1961–82

	Gross investment			
Accounting year	Current dollars (million)	Constant 1981 dollars (million)	As a proportion of total funds employed (percent)	Constant 1981 dollars per ton of copper produced
1961	22	76		134
1962	25	87		153
1963	18	63		115
1964	14	48		83
1965	60	203		315
1966	53	170		248
1967	46	145		247
1968	22	74		120
1969	45	151		229
1970	69	208	10.9	275
1971	104	291	13.7	391
1972	111	282	12.6	437
1973	147	313	13.4	435
1974	158	269	13.0	390
1975	164	246	11.0	353
1976	132	194	7.5	291
1977	105	142	6.9	206
1978	123	141	9.5	218
1979	96	99	5.7	159
1980	110	105	6.4	196
1981	200	200	10.8	340
1982	273	272	15.9	459

Source: Zambian Consolidated Copper Mines, Ltd.

data on capital funds used do not go beyond 1970. To get an impression of relative investment in the early 1960s, column 4 provides a second measure: annual gross investment in constant dollars per ton of copper produced.

A study of columns 3 and 4 suggests that investments were quite low through 1964, reasonably high in 1965–70 and 1976–80, and very high in 1971–75 and 1981–82.

A number of factors determine the volume of investment. The most important, both in private and publicly owned enterprises is probably the level of profits. The 1965 upward shift in the investments in Zambian copper is undoubtedly explained in the main by the sharp upward movement in copper prices in mid-1964, even though part of it might have been due to adjustments required by the border closure with Rhodesia. A similar coincidental movement between prices and investment levels, but in a downward direction, can be observed in the mid-1970s.

Two factors in combination probably explain why investment increased in the early 1980s despite continued low copper prices and profit levels. First, merely maintaining production of copper may have required new investment, given the deficient investment levels in the latter half of the 1970s. A second explanation is that investment reaction lagged the exploding cobalt prices in 1979–80. In fact, a substantial proportion of the overall investment expenditure in 1981 and 1982 was devoted to expanding cobalt capacity.[15]

Finally, the upward shift in investment about the time of nationalization can plausibly be seen as a consequence of the disappearance of political risk, which until then may have constrained the willingness of private foreign owners to commit investment funds.

The Zambian Copper Industry Under State Ownership

Would a private multinational enterprise have behaved more flexibly with regard to capacity utilization than did the state-owned Zambian copper industry? In only one instance did a clear-cut reduction occur in utilization of the technical capacity to produce copper in Zambia since nationalization. This happened in 1975, in response to the CIPEC decision to cut production 15 percent from the preceding year to reverse the very sharp price fall that occurred in the latter half of 1974. The implementation of the decision was not particularly effective. In 1975, Zambia's output was 3 percent lower than in 1974. The predominantly state-owned industries in Chile and Zaire, two other CIPEC members, reduced output by 8 percent and 1 percent, respectively. The privately owned copper producers in the United States, Canada, and Australia responded individually to the price fall, without national or international collaboration, by cutting their 1975 outputs by an average of 11 percent and 13 percent, respectively.

Production in Zambia recovered in 1976, and full capacity utilization has been maintained since that time. The declining output in the late 1970s was not due to restricted capacity utilization but to a reduction in technical production capacity. Even in the latter half of 1982, when real prices fell to the lowest levels recorded since the Great Depression and 25 percent of Western World mine capacity had been moth-balled,[16] the Zambian industry continued to produce at the maximum technically feasible rates. The low level of variable costs, explained to a considerable degree by the fixed nature of labor costs, provides an economic rationale for the Zambian industry's behavior.

15. Private communication with P. Crowson of RTZ.

16. Anthony Bird Associates, "Copper Analysis," London, June 1983.

Although the hypothesis that rates of capacity utilization are less flexible among state-owned than among private enterprises is certainly confirmed by comparing Zambia's copper industry with private firms in North America or Australia, one can only speculate about the Zambian industry's behavior had the copper firms remained in private foreign hands. Given the industry's overwhelming importance to the national economy, the government might have introduced labor legislation making labor virtually a fixed cost also to a private copper producer. Alternatively, to avoid the social and political strain of capacity shut-down, the government could have offered selective subsidies to the owners to reduce the risk of impending mine closures. The South African government used selective subsidies and tax exemptions precisely for this purpose in the 1980s to keep the privately owned Anglo-American gold mines in uninterrupted operation when gold prices fell.[17] Similar government action might well have coerced or induced a privately owned copper industry in Zambia to maintain production at full capacity levels throughout the period of low prices.

The figures on capital expenditure presented in the preceding section do lend some support to the hypothesis that investments will be higher under state ownership than under private foreign ownership in a situation in which the foreign owners feel exposed to political risk. The importance of this factor is difficult to determine, however, with the data that are available. In the Zambian case, copper prices certainly appear to be a far more important determinant of investment levels than the distinction between private foreign and public national ownership.

Whatever conclusions are drawn about investment, the 40 percent decline in Zambia's share of world copper production in the period of state ownership support the hypothesis that state-owned mineral enterprise is not the successful and aggressive institution that would threaten the survival of the private mineral industry. But then, an equally plausible view would be that this hypothesis remains unresolved, because ZCCM's performance is still suffering from immaturity and lack of experience.

The analyses of the preceding pages are also consonant with the hypothesis that takeovers by state enterprise involve substantial start-up and learning costs, possibly for quite an extended period of time. The widely admitted loss of efficiency due to Zambianization, as well as the failure to expand output in the years following nationalization, despite very substantial increases of both capital and labor, can reasonably be interpreted as start-up costs.

Higher cost levels in state enterprise are not necessarily a temporary phe-

17. Interview with C. Belshaw and P. Freeman of ZCCM.

nomenon, according to the literature on the subject. The mixture and ambiguity of goals in state-owned industries often permanently reduce pressure to minimize costs, as compared to private profit-maximizing units. The management of the Zambian copper industry strongly emphasizes employment and providing social amenities to the employees and to the local communities. Logically, profits should have been higher if the industry had not taken these social objectives upon itself. Nothing is necessarily antisocial about a pure profit-maximizing goal for the Zambian copper operations. The government could spend the higher profits to promote social development in the way it deems most useful. In fact, the copper industry does not appear to be particularly appropriate for employment creation, given its high capital intensity. Yet, it continues to be entrusted with (or takes upon itself) social objectives apart from profit maximization. In one executive's opinion, the reason for this pursuit of broader goals is that the industry may be a more efficient instrument than the government itself in accomplishing some of the social objectives.

Thus, in Zambia, state ownership has raised the cost of producing copper (1) temporarily, because of the costs of setting up the state enterprise and of allowing it to gain experience; and (2) permanently, because the wider goals reduce the importance of profit maximization and cost minimization. Although the features leading logically to higher cost levels can all be inferred from the earlier analyses, it is not possible to show in a clear-cut way that costs have actually increased since the government took a majority ownership position in the industry. The available cost data are rough and incomplete. Furthermore, the extended period of high copper prices just preceding nationalization might have led to a substantial organizational slack that would have been suppressed in the low-price period, had the private owners remained.

Nationalization in Zambia did not cause any important rupture of vertical integration chains. Most of the copper ore output was already being processed into refined copper within the country at the time of nationalization. Hence, the hypothesis about state enterprise establishment reducing vertical integration and expanding the arms-length mineral markets is not relevant in the Zambian case.

Even though four of the hypotheses—(a) lesser flexibility in capacity utilization, (b) no survival threat to private industry, (c) large start-up costs, and (d) permanently lower cost efficiency—all receive reasonable support from the figures and analyses of the Zambian case, the material available does not allow definitive tests of these hypotheses. (See appendix 6.A for a further caveat about the hypothesized distinctions between state and private enterprise.)

Appendix 6.A
A Private Enterprise Comparison

A further caveat to the hypothesized distinctions between state and private enterprise is provided by comparing the progress of the Zambian copper industry in the period of state ownership with that of Phelps Dodge, a leading U.S. copper producer. The reason for choosing Phelps Dodge among U.S. copper-producing firms is that Anaconda and Kennecott have merged with petroleum corporations, while in other potential candidates, copper accounts for a relatively limited share of total turnover (see T.T. Tomimatsu, *The U.S. Copper Industry: A Perspective of Financial Health*, U.S. Bureau of Mines, IC 8836, 1980, table 9).

In many readers' views, the proposed comparison may seem far-fetched. The two firms differ considerably from each other in many respects. Most importantly, while copper mining, smelting, and refining dominate completely in the Zambian industry, almost half of Phelps Dodge's sales stem from manufacturing. Yet, as appears in table 6.A-1, the similarities in terms of sales and shareholders' funds are striking.

The period under scrutiny is divided in the table into the high-price period, ending in 1974, and the low-price period covering the subsequent years. If it is true that efficiency in the state-owned copper industry in Zambia has suffered from the temporary start-up costs following national takeover of equity and management and from the permanently reduced pressure to minimize costs, then the Zambian firms should have experienced a more severe relative profit deterioration than Phelps Dodge from the early to the late 1970s.

A first scrutiny of the figures does indeed suggest that this was the case. Zambian sales increased much less and the pretax profit deteriorated much more than for Phelps Dodge. In the early period, the Zambian industry's pretax return on shareholders' funds was twice as high as for Phelps Dodge, but in the latter years, the Zambian copper companies ended up with a return on equity much below that recorded by the U.S. firm.

The argument so far has abstracted from the fact that Zambian copper is sold at LME quotations, while that of Phelps Dodge is typically transacted at domestic U.S. prices, over which producers have had considerable leverage. The average levels for the LME prices and for the U.S. producer delivered prices, as reported by *Metals Week* (reproduced in Phelps Dodge's Annual Report), are given in the table. Measured in real terms, the LME price has experienced a much sharper deterioration than the U.S. price.

The three bottom rows of the table are adjusted for the difference in the

Table 6.A–1. A Comparison of the Zambian Copper Industry with Phelps Dodge

Item	Zambian Copper Industry (accounting years ending March)		Phelps Dodge (calendar years)	
	1971–75	1976–82	1970–74	1975–81
Shareholders' funds at the end of each period (million dollars)	1,113	1,033	892	1,137
Sales (million dollars)	5,280	7,511	4,174	7,850
Profit before tax (million dollars)	1,965	184	755	476
Average annual pretax profit (% of shareholders' funds at end of period)	35	3	17	6
Tax payments (million dollars)	954	24	254	81
Tax payments as % of profit	49	13	34	17
Profit after tax (million dollars)	1,011	160	501	395
Average annual aftertax profit (% of shareholders' funds at end of period)	18	2	11	5
Average price, LME (Zambia) and U.S. producer delivered price (Phelps Dodge) cts/lb				
(current dollars)	67.2	72.7	59.0	77.4
(1981 dollars)	163.0	87.3	145.8	93.5
Hypothetical figures for Zambian industry, assuming sales at U.S. prices:				
Sales (million dollars)	4,636	7,997	—	—
Profit before tax (million dollars)	1,320	670	—	—
Average annual profits before tax for the period (% of shareholders' funds at end of period)	24	9	—	—

Sources: Zambian Consolidated Copper Mines, Ltd.; Phelps Dodge Annual Reports 1979 and 1981.

prices received, by recalculating the Zambian sales and pretax profits, on the hypothetical assumption that all costs remained as they were but that all copper sales had been made at the average prices received by U.S. producers. The picture that then emerges suggests quite similar sales growth for the subjects under study and a reduction by two-thirds in the before-tax rate of profit from the high-price to the low-price period for the Zambian industry, as well as for Phelps Dodge.

The comparison may be criticized on the ground that the circumstances under which Phelps Dodge operated in the period surveyed have not been analyzed at all. The profitability of Phelps Dodge could have been adversely affected by factors that it could not control. ZCCM, however, has also been hit by adverse factors unrelated to state ownership such as the substantial fall in ore grades mined and the inadequate availability of foreign exchange in Zambia.

The results of the corporate comparison must be regarded as highly tentative, but they do cast some doubt on the contention that state ownership regularly results in higher costs and lower efficiency.

Appendix 6.B
Zambian Interviews

National Commission for Development Planning

Mr. L. Nkhata, Investment Policy Department
Mr. G. Chivunga, Mineral Economist, Sectoral Planning Department

Ministry of Mines

Dr. E.H.B. Mwanagonze, Permanent Secretary

ZIMCO

Mr. J. Mapoma, Chairman

Memaco

Mr. L.C. Mutakasha, Managing Director

University of Zambia

Dr. C. Chibaye, Dean Humanities Department
Dr. C. Mpaisha, Professor Public Administration
Dr. M. Kaniki, Professor Economic History

Other

Mr. D.A.R. Phiri, Zambian Ambassador to Sweden, former Managing Director of Roan Consolidated Mines Ltd.

ZCCM

Lusaka Head Office:

Mr. F.H. Kaunda, Chairman and Chief Executive
Mr. J.D. Chileshe, Director Industrial Relations
Mr. M.D. Sichula, Director Manpower Development and Planning
Mr. J.F. Godsall, Director Corporate Planning
Mr. B.G. Moyo, Director Administration
Dr. E. Koloko, Deputy Director Corporate Planning
Mr. C. Belshaw, Consulting Mining Engineer
Dr. P. Freeman, Consulting Geologist

Copperbelt:

Mr. J.R. Hoatson, General Manager Nchanga Division
Mr. I.S. Blair, General Manager, Mufulira Division
Mr. W.B. Eastwood, General Manager, Nkana Division
Mr. J.C. Vergeer, General Manager, Centralized Services Division
Mr. D.A. Lendrum, Mine Manager, Nkana Division
Mr. E. McGuiness, Geologist, Nkana Division
Mr. A. Nelson, Accounts Manager, Centralized Services Division
Mr. L. Hewagama, Chief Accountant, Centralized Services Division
Mr. G. Cutler, Manager Purchases, Centralized Services Division

III

CONCLUSIONS

7

Conclusions

Although bits and pieces of evidence to support the claim of a fast-growing state-owned mineral sector abound, I was unable to find any systematic and reasonably general evidence to support the premise that state enterprise has become an important factor in global supply, and particularly so in developing countries' mineral industries. The detailed data presented in chapter 2 constitute my own attempt to measure the extent and growth of state ownership in the mineral industries.

The features that characterize state enterprises are the result of state control over corporate activity. The degree of control, however, is not always uniformly related to the extent of state ownership. Given the difficulties in defining the concept and establishing the degree of control, empirical measurements of the size of the state enterprise sector are invariably based on the magnitude of the governments' equity ownership. Even with this simplification, the prevalence of state enterprise can be measured in at least three different ways: as (a) production capacity in which the government has significant equity ownership, or (b) production capacity in which government has majority ownership, or (c) production capacity proportional to the government equity holding in each production unit. Table 2-1 demonstrates that the differences among the three measures can be quite substantial.

In the Western World, state enterprise in the mineral industries is much more prevalent in developing than in industrialized countries, even though the latter have significant ownership positions in some minerals—for example, aluminum and iron ore/steel. When the extent of state enterprise is measured by the capacity proportional to the government equity holding, the figures suggest that roughly one-half of the nonfuel mineral industries in developing countries are government owned. The figure falls to about one-third when the entire Western World is considered. These measurements clearly indicate that state-owned enterprises are a very important factor in world mineral markets.

Public ownership of the mineral industries on a large scale in the Western World is a new phenomenon. As recently as the mid-1950s, state involvement in this sector outside the Socialist countries was insignificant. At that time, the mineral industries of Africa, Asia, and Latin America were completely dominated by privately owned multinationals from the leading industrialized market economies. The emergence and growth of state enterprise has been prompted by a variety of factors, but a major one has been the political independence and economic emancipation of the Third World. Between 1960 and 1975, a substantial proportion of the total mineral industry in developing countries was nationalized. The motivations for takeover included claims that the colonial arrangements with the private multinationals were of little benefit to the newly independent nations; convictions that economic independence impinged upon government control of key economic sectors like minerals; and widespread socialist political philosophies, according to which the major means of production should be publicly owned. Viewed in this light, the emergence of the state-owned mineral sector in developing countries can to a large extent be seen as a one-time adjustment in the wake of economic independence.

My analysis suggests that the phase of fast growth may now have ended and that the proportion of state ownership in Western World mineral activity will not expand much above current levels. Several factors support this claim. First, the adjustment to rectify the former unfavorable colonial conditions that prevailed in developing countries has probably come to an end—the governments have already absorbed the most conspicuous foreign ownership positions. Furthermore, the relationships between governments of developing countries and multinational mining firms in the 1980s appear much less conflict-ridden than ten or twenty years earlier.

The claim that state-owned enterprises in any given mineral will expand their share of the overall market in the absence of further nationalizations has no empirical support. The governments are anxious to maintain their involvements in the mineral activities but not necessarily to expand them above present levels. New projects are commonly launched as joint ventures with one or several multinationals, with the government holding an important equity position, but not necessarily a majority one. Also, in the 1980s, social-

and 1960s, immediately after a majority of the countries in Africa and Asia gained independence. Over time, ideology has come to play a lesser role than the ability to deliver the desired results when choosing ownership forms in the mineral industries. A third factor is the increasing sophistication of private sector entrepreneurship and management found within the developing countries themselves. The second and third factors in combination can be expected to lead to a greater role for private national enterprises in the continued expansion of mineral sector activity in developing countries whose governments resist foreign ownership in base industries.

Public ownership positions in mineral industries in industrialized and developing countries are not likely to be dismantled; but even though some limited expansion in the Western World state-owned share may take place as a result of the continuing shift of mining and mineral processing towards developing countries, the arguments suggest that the fast growth of the state-owned share is now over.[1] The tentative conclusion is that in the 1980s and 1990s the expansion of state enterprise in Western World mineral industries will not be much different from the overall growth of the mineral sector. Hence, the functioning of the international mineral markets will continue to be conditioned by the presence of an increasingly mature group of state-owned mineral producers representing a substantial but relatively static share of world supply.

The Market Impact of State Enterprise Proliferation

An important conclusion that emerges from the analyses of chapters 2 and 3 is the blurred nature of the state enterprise concept. State enterprises come in many shapes with regard to objectives and operational modes. The emergence of state mineral enterprises in many countries has been coincidental with important changes in their social environments. These changes have influenced the behavior of the private firms. For these reasons, the distinction between the state-owned and private multinational firms will be much less sharp than would appear from the ideal models discussed in part I.

The five hypotheses about the market impact of state enterprise formulated at the end of chapter 3 were derived from insights gained from the economics literature on state enterprise, from generalizations based on wide but scattered empirical observations, and from general economic reasoning and logical deductions. Formal proof of the validity of the hypotheses would require additional research and a far greater systematic data base than has been

1. For evidence of such a shift see M. Radetzki, "Has political risk scared mineral investments away from developing countries?", *World Development*, no. 1, 1982.

possible to establish in this study. Even if such a data base were available, confirmation of the hypotheses might not be possible.

Even though the present informal and qualitative test of the hypotheses does not permit the formulation of general conclusions, the outcome of the exercise is of value in demonstrating whether or not the plausible suppositions contained in the hypotheses are confirmed or contradicted by the three case studies.

The first hypothesis is that the establishment of state mineral enterprises has temporarily led to lower output levels and higher costs and prices than would have prevailed with the old ownership patterns.

The start-up and learning costs after nationalization are evident in all three case studies but are most apparent in the Indonesian tin industry (chapter 4). There, the difficulties of the inexperienced national managers were compounded by the political turmoil during the Soekarno years but stretched well beyond that period. All three industries suffered a decline in world market shares. Production and capacity utilization fell, with an ensuing rise in costs in the years after state takeover. In the Venezuelan example, this development is explained primarily by the rupture of vertical integration; in the Indonesian and Zambian cases, by managerial deficiencies. The extended inability to expand capacity is also clearly apparent in the Indonesian and Zambian studies. In Venezuela too, capacity stagnated or contracted, but depressed world market conditions provide a plausible additional reason for this development.

Thus, the case studies appear to support the hypothesis that the establishment of state mineral enterprises entails start-up costs in the form of lower output and higher costs than would have prevailed with unchanged ownership arrangements. The evidence contained in the case studies, however, is unsuitable for testing the logical deduction that mineral prices will temporarily be higher as a result of the disruption and efficiency problems following widespread nationalizations.

The second hypothesis is that mineral production costs in state-owned enterprises will be permanently higher than in private firms under similar circumstances because of the costs state-owned firms incur in their pursuit of nonprofit objectives and because of the lesser cost-minimization pressures in such firms.

All three case studies clearly indicate that the governments used the state-owned mineral firms heavily to promote a variety of social goals. As a consequence, these firms did incur significant costs that could have been avoided by private enterprise in the respective countries. Given these broader goals, however, the case study material has not demonstrated that costs were higher than in private enterprise because of less pressure on management to minimize costs.

Thus, the case studies do support the hypothesis that mineral production costs in state-owned enterprises will be higher as a result of the typical pursuit of nonprofit objectives, but the evidence of higher cost due to permanent inefficiency is unclear.

The third hypothesis is that the emergence of state enterprises has stabilized output over the business cycle and hence destabilized medium-run prices in mineral markets.

Some indirect support for the hypothesis follows from the observation that the state enterprises studied have very little opportunity to vary their labor force and tend to regard labor costs as fixed. The direct evidence from the case studies in support of the hypothesis is inconclusive or weak. The Indonesian tin deposits are low cost, and tin prices have been so high throughout the period of state ownership that full capacity utilization would in all probability have been rational even if the industry had been operated by private interests. The Venezuelan iron ore producer did reduce production after 1975, but this development was more due to the market disruptions following nationalization than to a conscious measure aimed at maximizing profits in the wake of falling prices. Only the Zambian case provides some support for the thesis that state firms are less flexible than private ones in terms of downward adjustments of production. The case study shows that Zambian copper output in 1975 fell much less than in several countries with predominantly private ownership.

Thus, the hypothesis about state enterprise growth leading to greater output stability and increased price instability in mineral markets over the business cycle is not confirmed, but neither is it disproved by the material contained in the case studies.

The fourth hypothesis is that the emergence of state mineral enterprises has permanently reduced the extent of vertical integration, substantially widening and stabilizing some formerly thin arms-length markets such as those for iron ore and bauxite.

The Zambian takeover did not involve any rupture of international vertical integration, but the Indonesian and Venezuelan cases did. Some vertical integration was subsequently reestablished in these two countries when Indonesia started smelting tin and Venezuela producing steel. The current extent of vertical integration of Indonesian tin and Venezuelan iron ore is less than before state takeover, however. Also, the shift of Venezuela's sales from captive markets in the United States to arms-length arrangements in Europe did contribute, albeit marginally, to the widening of noncaptive international iron ore trade.

Thus, the findings of the three case studies do provide support for the hypothesis that the growth of state enterprise has reduced vertical integration in international mineral markets and expanded noncaptive trade. But the material contained in the studies is not conducive to testing the supposition about an ensuing increase of price stability in arms-length markets.

The fifth and last hypothesis formulated in chapter 3 is that state enterprises do not threaten the survival of the private mineral industry worldwide.

The evidence is that market shares fell in Venezuelan iron ore and Zambian copper; that investment was small and no concessional financial infusions were made in the Indonesian and Venezuelan state firms, despite those countries' swelling government budgets through the 1970s; that the Zambian industry was unable to increase output in response to higher capital and labor inputs; and that state-owned and private multinational tin producers attained peaceful coexistence in Indonesia. These observations are contrary to the view of state mineral enterprises as aggressive institutions that threaten the existence of private mineral endeavors.

Both publicly owned and private enterprises may sometimes exhibit very aggressive market expansion behavior. Ferrominera pursued lethargic investment and marketing policies, but the state-owned CVRD of Brazil pushed energetically to increase its market share. The iron ore shipments under CVRD's control increased from 3.9 percent of world output in 1971 to 9.3 percent in 1982, or by a factor of 2.4.[2] Such behavior is not exclusive to state-owned enterprises but can also be found among individual private multinational corporations. For example, mine production of copper under the control of Rio Tinto Zinc, a UK multinational resource company, rose from 1.7 percent of world output in 1971 to 4.7 percent in 1982, or by a factor of 2.8.[3]

These three case studies reasonably confirm the hypothesis that the emergence of a state-owned enterprise does not constitute a survival threat to private multinational mining firms.

At this point, a small digression may be in place. Most of the laments private miners expressed from 1982 through 1984 about the threat state-owned enterprises pose originated in North America and concerned copper. That

2. *Mining Annual Review*, 1973 and 1983.

3. RTZ Annual Reports, 1972 and 1982.

period recorded a severe trough in copper consumption[4] and the lowest real copper prices since 1950.[5] North American producers were much more severely hit than producers in other parts of the world because of the appreciation of the U.S. dollar (plus 27 percent in terms of special drawing rights between 1980 and 1984) and because discrete devaluations improved the competitiveness of the copper industries in many other countries. The new exchange rates, in my opinion, were the major threat to the survival of individual North American copper-producing firms in the early 1980s, not state enterprise or developing countries. For example, primarily as a result of the depreciation of the Swedish currency, the Swedish mining company, Boliden, has been able to cover operating costs at full capacity utilization in 1982 and to reap net profits from its copper operations in 1983, a feat not achieved by many North American copper producers.

It is true that improvement in relative competitiveness in the copper market has been greatest in developing copper-dependent countries like Zaire and Peru, which undertook especially large devaluations. These countries, however, had especially large imbalances in their current accounts, caused by the declining world demand for copper and also by such factors as high internal inflation and the sharp rise in international interest rates in the early 1980s. Substantial devaluations were required to correct the current account imbalances.

By lowering local costs, the devaluations have improved the international competitiveness of copper from many developing countries and have made full capacity utilization throughout the recession an economically rational policy for both private and state-owned producers. Subsidized credit from national governments and international agencies have played a very small role in this context in comparison with the impact of exchange rate changes.

The corollary of these developments has been reduced international competitiveness of copper producers in North America. Unless the relative costs of copper production change again, the result will be that the high-cost North American producers will be forced out of business altogether. The president of Kennecott Corporation estimates that about 12 percent of existing U.S. copper mining capacity in 1983 is not viable in the long run.[6]

In summary, then, it is true that the high-cost segments of the mineral industries in some industrialized countries have been exposed to a survival

4. *Metals Analysis and Outlook*, no. 18, fourth quarter, 1983.

5. World Bank, *Commodity Trade and Price Trends*, 1983-84 edition; *CIPEC Quarterly Review*, fourth quarter, 1983.

6. *CIPEC Quarterly Review*, fourth quarter, 1983.

threat in the early 1980s. In my opinion, this threat has been primarily caused by substantial changes in relative competitiveness among all mineral producers, resulting from the large-scale adjustments of rates of exchange. Thus, the threat has little relationship to the emergence of state-owned mineral enterprises or to the purported ease of access such enterprises have to concessional financing.

Conclusion

The hypotheses formulated in this study suggest that state enterprises will indeed affect mineral markets in important ways. The empirical evidence collected in the course of the present work, however, cannot unambiguously confirm these hypotheses. Hence, few definitive statements can be made about the implications for international mineral markets of the increasingly important ownership positions of governments in mineral industries.

This in itself is an important conclusion. The many claims made by private industry, governments, or international organizations that state ownership has fundamentally changed the rules of the game in one respect or another still remain to be proven, and the present study illustrates that this is not easily done.

Additional collection of empirical data to verify the hypotheses would appear to be a fruitful task. Such work could be pursued along two paths. The first would be to undertake further case studies of individual state-owned enterprises in order to widen the sample against which all the hypotheses could be tested. Useful additions might include studies of state enterprises, such as CVRD in Brazil and Codelco in Chile that have aggressively expanded their market shares; additional minerals, such as bauxite and nickel, with market circumstances that differ from the ones considered; and enterprises in which the government holds less than 50 percent of the equity.

The second path might be to conduct a more intensive search for data on a wide basis concerning a single aspect of market conditions in the presence of state enterprise. Examples might include output growth after state takeover, relative flexibility of capacity utilization in state-owned and private enterprises, or the extent of ultimate change in vertical integration resulting from the growth of state ownership.

Index